SPECIAL PAPERS IN PALAEONTOLOGY NO. 66

ANGIOSPERM WOODS FROM BRITISH LOWER CRETACEOUS AND PALAEOGENE DEPOSITS

BY

MARK CRAWLEY

with 13 plates, 16 tables and 18 text-figures

THE PALAEONTOLOGICAL ASSOCIATION
LONDON

October 2001

CONTENTS

ABSTRACT. Four of the five putative British Lower Cretaceous angiosperm woods *Aptiana, Cantia, Hythia, Sabulia* and *Woburnia* (Stopes 1912, 1915) are re-evaluated. *Aptiana radiata* Stopes, 1912 is accepted as Lower Cretaceous (Aptian/Albian) and is, therefore, regarded as the only valid British Cretaceous angiosperm wood. New material of *Cantia, Hythia* and *Sabulia* has allowed an original provenance of Palaeogene for all specimens representing these taxa. They also show similarities to Betulaceae (*Cantia*), Icacinaceae, Platanaceae or Fagaceae (*Hythia*) and Lauraceae (*Sabulia*). Fifteen new species are described: *Anacardioxylon maidstonense, Apocynoxylon? oldhavenense, A. sapotaceoides, Canarioxylon lewisii, Castanoxylon philpii, Dryoxylon calodendrumoides, Entandrophragminium lewisii, Euphorbioxylon hernense, Flacourtioxylon oldhavenense, Ilicoxylon? prestwichii, Meliaceoxylon collinsonae, Paraphyllanthoxylon chievleyense, Polyalthioxylon oldhavenense, Tetrapleuroxylon oldhavenense* and *Tilioxylon lueheaformis*. Three new combinations of Palaeogene wood are also described. The new species and new combinations show feature sets found in Recent Anacardiaceae, Annonaceae, Aquifoliaceae, Apocynaceae, Burseraceae, Caesalpinaceae, Euphorbiaceae, Fagaceae, Flacourtiaceae, Hamamelidaceae, Meliaceae, Lauraceae, Lecythidaceae, Sapotaceae and Tiliaceae. All British Palaeogene material is reviewed for wood and tree evolution, palaeobiology, palaeobiogeography and palaeoclimatology. The anatomical results show increased diversity by the latest Palaeocene, including the oldest known wood with spiral thickening of the vessels, and support a trend of increasingly warm temperatures with less seasonality and structures more typical of Recent tropical regions by Late Palaeocene/Early Eocene times in the British area.

FOR 75 years Marie Stopes' angiosperm woods have sat uncomfortably in the Aptian. In her original study, 'Petrifactions of the earliest European angiosperms' (1912), she described three dicotyledonous woods from the Aptian Lower Greensand: *Aptiana radiata, Sabulia scottii* and *Woburnia porosa*. This study was followed (Stopes 1915) by descriptions of two additional taxa: *Cantia arborescens* and *Hythia elgarii*. Stopes (1912, p. 78) acknowledged the problems caused by these specimens, which seemed 'advanced' despite their early age, but she seemed convinced of their Aptian age. She stated 'While I have not the same absolute guarantee that these specimens are from the deposits whence they are described which I should have had I found them myself, yet there is nothing to cause one to doubt the correctness of their registration. There are, moreover, a number of points of internal evidence in the specimens themselves, which confirm their allocation and appear to be entirely convincing when taken in conjunction with the Museum labels.'

Page (1981) expressed doubts about the age assignment, concluding that the specimens may have been erratics. Hughes (1961) observed that no further angiosperm woods had since been found at any of the Greensand localities and concluded that confirmation by finding a new specimen of any angiosperm wood in these beds was needed. In 1960 Dr Raymond Casey examined the specimens at the request of Prof. T. M. Harris [unpublished report, British Geological Survey (BGS), 1960], concluding that the three specimens with adherent matrix (*Aptiana, Cantia, Hythia*) did, indeed, appear to be from the Lower Greensand. Not surprisingly, therefore, the specimens have remained controversial. Two of the woods show specialization that appears inconsistent with their putative Early Cretaceous age: *Sabulia* (aliform and banded parenchyma) and *Woburnia* (extremely large pores, short to medium vessel elements, vasicentric tracheids and longitudinal canals; Page 1981). It is important to try and resolve the age as records of angiosperm woods from Aptian and Albian deposits are rare (Page 1981; Serlin 1982; Thayn *et al.* 1983, 1985).

Four of the five putative British Lower Cretaceous angiosperm woods, *Aptiana, Cantia, Hythia, Sabulia* and *Woburnia* (Stopes 1912, 1915), are re-evaluated here. *Aptiana radiata* Stopes 1912 is accepted as Lower Cretaceous (Aptian/Albian) and is, therefore, regarded as the only valid British Cretaceous angiosperm wood macrofossil. New material of *Cantia, Hythia* and *Sabulia* discovered in the collections of The Natural History Museum between 1989 and 1990 indicates a Palaeogene age for these taxa. They are also assigned to families: Betulaceae (*Cantia*), Icacinaceae (*Hythia*) and Lauraceae (*Sabulia*). Further details can be found below in the introductory sections for the Lower Greensand, Thanet Sand Formation and Oldhaven Beds.

Dicotyledonous woods are an important component of several British Palaeogene floras (Thanet Sand Formation, Reading Formation, Oldhaven Beds and London Clay Formation). They may help to shed light on the composition and characteristics of past floral communities. Fossil wood can also be of

[Special Papers in Palaeontology, 66, 2001, pp. 1–100, 13 pls]

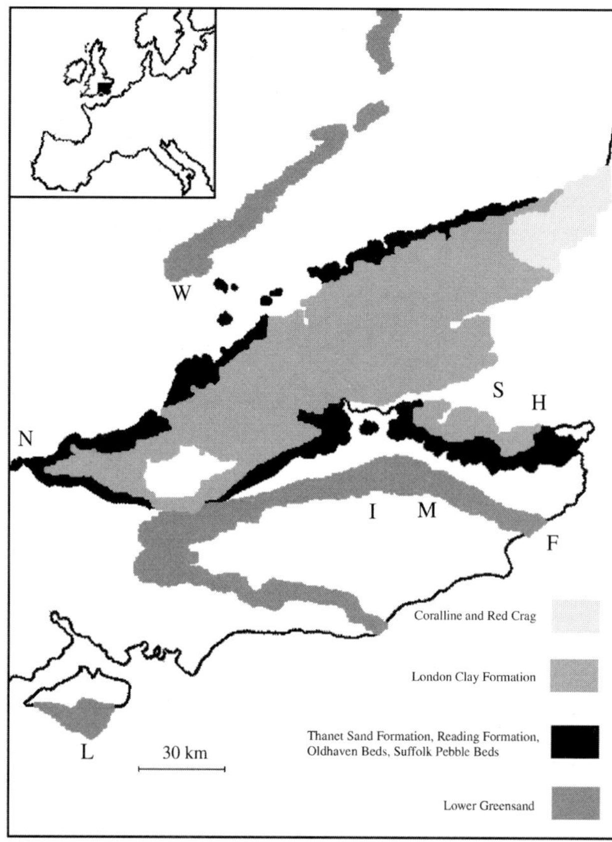

TEXT-FIG. 1. Distribution of sedimentary units in southern Britain; F, Folkestone; H, Herne Bay; I, Ightham; L, Luccomb Chine; M, Maidstone; N, Newbury; S, Isle of Sheppey; W, Woburn Sands (based on Collinson 1983; Harland *et al.* 1989; and British Geological Survey, $\frac{1}{4}$" Series, sheet 2).

value in palaeoclimate interpretation (Creber and Chaloner 1984; Wheeler and Baas 1993; Wiemann *et al.* 1998, 1999; Creber and Francis 1999). Also, the material has allowed this re-evaluation of taxonomy and age of four of Stopes' earliest angiosperm macrofossils. The aim of this study is to describe dicotyledonous woods in museum collections as well as material collected recently. All of the material is from south-east England and ranges in age from Early Cretaceous, about 120 Ma, to Early Eocene, about 55 Ma (see Text-fig. 1). Lithostratigraphical classification of the Palaeogene strata of the London Basin and East Anglia is that of Ellison *et al.* (1994). The bulk of this work is concerned with wood assemblages from the Thanet Sand Formation (Palaeocene), Reading Formation (Palaeocene), Suffolk Pebble Beds and Oldhaven Beds (uppermost Palaeocene/lowermost Eocene; M. E. Collinson, pers. comm. 1995) and London Clay Formation (Lower Eocene). Importantly, Palaeocene vegetation is among the least well known (Wolfe 1985); for example, only 37 wood records are documented worldwide, most of which do not have secure age determinations (Wheeler and Baas 1991). The text is arranged sequentially by age and formation, and each wood flora is separately described within these sections. All previously described silicified and calcitized angiosperm wood from the Inner Hebrides (Crawley 1989, part of the Brito-Arctic Igneous Province Flora of Boulter and Manum 1989), and the Palaeogene (Crawley 1989; Brett 1956, 1960, 1966, 1972; Pearson 1987, 1989) are brought together in the conclusion.

MATERIAL AND METHODS

All of the specimens were studied by thin sections of transverse, tangential and radial surfaces (abbreviated in the text as TS, TLS and RLS) using a Leitz Ortholux compound microscope. The number of vessels per unit area (mm^2) was calculated by counting all individual vessels, including those in groups or clusters (Wheeler 1986). Percentages of solitary vessels per unit area (mm^2) were also calculated from total individual vessel counts. Quantitative ranges and means are based on at least 20 measurements where possible. Vessel element length includes the vessel 'tail' where visible. The term 'mean range' is used when more than one specimen has been measured and shows the range of the different means. The contents of the 'Anatomy of the Dicotyledons' by Metcalfe and Chalk (1950) were used as a basis for comparison with extant families unless otherwise specified. Computerized searches were conducted of extant and other fossil woods using the General Unknown Entry and Search System (GUESS) computerized database (Wheeler *et al.* 1986; Wheeler and Baas 1991). Descriptions herein include numbers, e.g. [**1**], relating to the International Association of Wood Anatomists' (IAWA) list of features for hardwood identification (Wheeler *et al.* 1989). This practise was started by Selmeier (1990) and is continued here as positive standardization for angiosperm wood data.

The specimens examined are in the palaeobotanical collections of the Department of Palaeontology, The Natural History Museum, London (NHM), and the geology collections of the Ipswich Museum, Suffolk (IM), Maidstone Museum and Art Gallery, Kent (MM), and the Sedgwick Museum, Cambridge (SM).

SYSTEMATIC PALAEONTOLOGY

Wood from the Lower Greensand

Stopes (1912, 1915) gave the age of this specimen as Aptian. A label on it stated Lower Greensand, with which she concurred, citing evidence from the surrounding matrix when compared with vouchered samples of known provenance. No locality information was with the specimen but Stopes thought that there was a strong likelihood of it being Luccomb Chine, Isle of Wight, where wood samples, albeit of gymnosperms, are not uncommon. Casey's examination of this specimen led him to conclude (unpublished BGS report, 1960) that the matrix was 'absolutely typical of some of the remanié beds in the Folkestone Beds division of the Lower Greensand. If handed to me for an opinion on horizon I would have said Luccomb Chine (near base of Folkestone Beds, top of Aptian) or Copt Point, Folkestone (top of Folkestone Beds, Sulphur Band, top of Lower Albian, taken as base of Gault by some authors). Both localities and horizons are replete with fossil wood.' Of all Stopes' specimens this is the only one with substantial adherent matrix and, in this case, it would seem to concur with Casey's view. The macro features of the wood are also identical to that of other woods from this horizon in the Lower Greensand, particularly the colour.

Class MAGNOLIOPSIDA Cronquist, Takhtajan and Zimmerman, 1966

Morphotaxon APTIANA Stopes, 1912

Type species. Aptiana radiata Stopes, 1912, Aptian, southern England.

Aptiana radiata Stopes, 1912, emend.

Plates 1–2; Table 1

1912 *Aptiana radiata* Stopes, p. 84, pl. 6, figs 1, 3–5; pl. 7, fig. 6; pl. 8, figs 10–11; text-figs 1–5.
1912 *Aptiana radiata* Stopes; Moll and Janssonius, p. 622.
1915 *Aptiana radiata* Stopes; Stopes, p. 284; text-figs 87–92.
1924 *Aptiana radiata* Stopes; Bailey, p. 448.
1924 *Aptiana radiata* Stopes; Scott, p. 54, fig. 7.

1931 *Aptiana radiata* Stopes; Edwards, p. 19.
1932 *Aptiana radiata* Stopes; Bancroft, pp. 356–357.
1971 *Aptiana radiata* Stopes; Pant and Kidwai, p. 252.

Holotype. NHM V.11517; slides V.11517a-i.

Locality and horizon. Unknown; inferred to be Luccomb Chine, Isle of Wight (Stopes 1912, 1915), or Copt Point, Folkestone, Kent; respectively Aptian, base Folkestone Beds, or Albian, Sulphur Band, top Folkestone Beds (see remarks by Casey noted above), Lower Greensand.

Emended diagnosis. Primary wood without marked 'bundles'. Secondary wood with vague growth rings marked by alternating zones of vessels with fibre-tracheids and fibre-tracheids with few or no vessels. Vessels are solitary and 11–60 μm in tangential diameter. Intervascular pitting alternate and in uniseriate files, bordered with a horizontal to oblique aperture of up to 3 μm. Vessel to ray pitting mainly alternate, bordered, round to elongate or gash-like, with apertures of 1–11 μm. Perforation plates simple and scalariform with up to 24 bars. Axial parenchyma is absent. Rays are uniseriate and multiseriate usually 3–6 cells wide with evidence of aggregation and fusion. Rays are heterogeneous, uniseriate, consisting of upright cells; multiseriate ray bodies composed of square and procumbent cells with a single upright marginal row. The multiseriate rays in the secondary phloem expand to funnel-shaped ends. The periderm consists of sclerotic islands in a reticulum of thin-walled cells.

Description. A silicified portion of axis, 360 mm in diameter consisting of secondary and primary wood with some periderm present (Pl. 1, fig. 1). Possible indistinct growth rings [**2?**] which microscopically are marked by few or absent vessels in the 'late-wood' (Carlquist 1988, Type 5). The fibre-tracheids show no real change in dimension or wall thickness across the 'ring' zone. Primary wood although present is poorly preserved.

Vessel elements. Alternating bands or zones of vessels with fibre-tracheids and fibre-tracheids with few or no vessels (Pl. 1, fig. 2; vessel bands top and bottom of figure). Semi-ring porous? [**4?**], exclusively solitary (Pl. 1, fig. 2) [**9**], apparent pairs are due to overlapping vessel element ends. Perforation plates are simple (Pl. 2, fig. 4) [**13**] and scalariform (Pl. 2, fig. 5) [**14**] with up to 24 bars [**17**]. Vessel to vessel pitting bordered, alternate, round to oval (Pl. 2, figs 2–3). Pit apertures oblique to almost vertical, about 3·5 μm [**24**] diameter reducing to 1·5 μm at vessel tail. Vessel-ray pitting is mainly bordered, alternate to irregular in distribution (Pl. 2, figs 6–7). Possible coalescent apertures were seen and some pits may be vestured [**?29**] (Pl. 2, fig. 6); individual pits are bordered, round to elongate and gash-like [**30–31**], with horizontal apertures of 1·4–10 μm; unilaterally compound pitting is present [**34**] (Pl. 2, fig. 7). Vessel to fibre pitting is a single vertical row of bordered pits (Pl. 2, fig. 1). Tangential diameter of vessel lumina range 11–60 μm, mean range 21–41 μm [**40**]; density 48–79 mm^2, mean 63 mm^2 [**49**]. Only a few vessel elements could be measured but a maximum length 810 μm was recorded [**54**].

Imperforate tracheary elements. Fibre-tracheids occur as ground tissue as well as in bands of just fibre-tracheids (Pl. 1, fig. 2: middle of figure). Small circular to oval bordered pits 3–4 μm in diameter [**61**] (Pl. 2, fig. 2: left); pit apertures oblique to vertical with pit pairs crossed.

Axial parenchyma. This feature was not seen [**75**].

Ray parenchyma. Rays are 1–6 cells wide [**97–98**]; uniseriate rays are abundant; multiseriate rays show evidence of aggregation [**101**] (Pl. 1, fig. 5), with the widest of these rays (10–13 seriate) appearing abruptly (Pl. 1, fig. 3), usually in the late wood. These rays can be very irregular and of short radial duration (Pl. 1, fig. 4). Stopes (1912) noticed an unusual feature of the rays, namely a complete disappearance or dwindling to only uniseriate as they pass radially outwards (Pl. 1, fig. 1). Multiseriate ray height range 484–1210 μm [**102**] (12–32 cells), mean 839 μm (22 cells); uniseriate ray height range 154–770 μm (2–7 cells), mean 427 μm (4 cells); multiseriate ray width range 44–110 μm

---

EXPLANATION OF PLATE 1

1–5. *Aptiana radiata* Stopes, 1912, emend., Lower Cretaceous, Aptian or Albian, Lower Greensand, southern England; The Natural History Museum, London, holotype, NHM V.11517. 1, TS, showing intermittent multiseriate rays; periderm is seen at the top of the figure, NHM V.11517r; ×25. 2, TS, growth ring showing zone lacking vessels, NHM V.11517r; ×100. 3–4, TS, very wide multiseriate rays with evidence of aggregation (compare with 2), NHM V.11517a; ×100. 5, TLS, aggregate multiseriate ray flanked by smaller multiseriate rays and uniseriate rays, NHM V. 11517j; ×100.

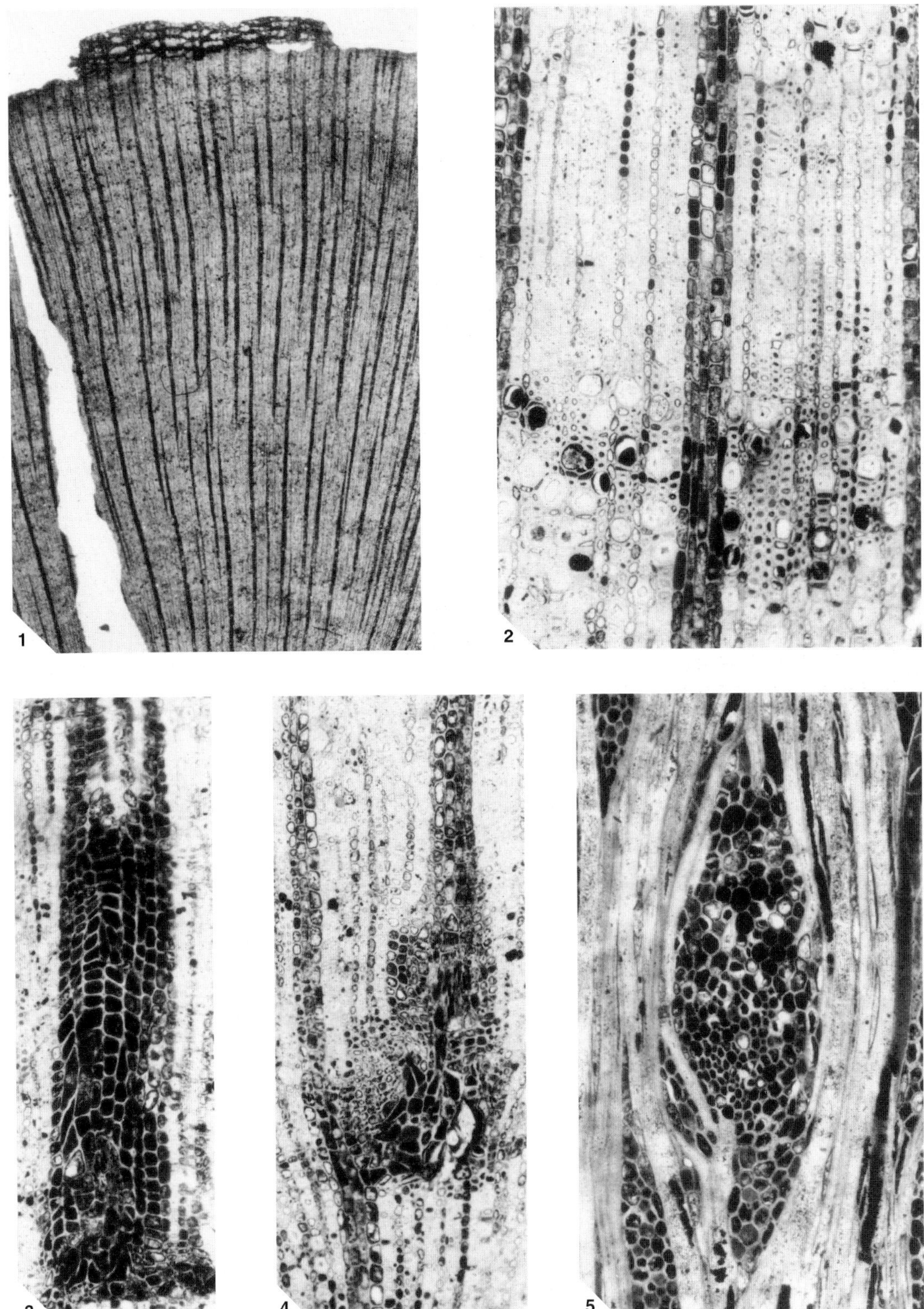

CRAWLEY, *Aptiana*

(3–6 cells), mean 83 μm (4–5 cells). Rays of two distinct sizes **[103]**, uniseriate and four or more seriate (Pl. 1, fig. 5); composition is heterogeneous, with usually one row of upright cells **[106]** and ray bodies composed of procumbent and square cells. Sheath cells are common **[110]** and are due to marginal fusion of uniseriate rays. Uniseriate rays consist entirely of upright cells. In transverse section a distinct chevron pattern of the marginal ray cells is seen in rays over five cells wide (Pl. 1, fig. 3); ray density range 10–22 mm^2, mean 15 mm^2 **[116]**.

Bark. This forms an incomplete sheath of about 1 mm thick (Pl. 1, fig. 1). It consists of islands of sclereids or sclerotic tissue in a reticulum of possible cork cells with dark contents. In TS the multiseriate rays become broader and funnel shaped within this zone.

Remarks on the wood anatomy. A feature of this wood is the arrangement of the vessels and fibre-tracheids. This was particularly well shown by Stopes (1915, text-fig. 89). Examination of TS thin sections shows that there are few to no vessels in bands or zones (up to 20 fibre-tracheids across) between alternating zones containing vessels with fibre-tracheids. Each zone or band is of similar width. This structure was confirmed by examination of RLS thin sections (NHMV.11517b and 11517q). This may be a type of semi-ring porous structure. Ring porosity is believed to be an adaptation to seasonal variation (Gilbert 1940). It may also be classified as a type of ring (Type 5 of Carlquist). Given the age of the specimen it may be neither. A study by Wheeler *et al.* (1995) shows that certain combinations of features common in the Cretaceous are rare in Tertiary or Recent woods. They argue that this may reflect hydraulic strategies that could be unique to the Cretaceous. Hence, I think it best to remain cautious about what this feature represents in *Aptiana*. Another feature of this wood is the possible presence of pit vestures. These structures are of limited occurrence today and are, therefore, of great diagnostic value (Metcalfe and Chalk 1950). This would be the earliest occurrence of this feature.

Comparison with Recent woods. Various hypotheses about the affinities of these woods have been made by Stopes (1912, 1915), Moll and Janssonius (1912) and Bailey (1924). They all considered the wood to be angiospermic but differed in their conclusions (see Edwards 1931 or Bancroft 1932 for summary). Stopes (1912, p. 90) mentioned in passing 'I have been told it resembled closely nearly every family ranging from the Gnetales on one hand to the Malvales on the other'. I was not able to find a convincing Recent match for this wood. Given the age of this specimen and the variation in attempts to classify it, a review of its anatomical features follows.

The secondary xylem of *Aptiana* has the following main features: zones of fibre-tracheids and vessels, which may be a type of semi-ring porous vessel arrangement or a type of growth-ring; solitary vessels; simple and scalariform perforation plates; possible pit vestures; fibre-tracheids; and heterogeneous ray structure. All of these features can be found in Recent angiospermic wood but not in any one genus and most in *Ephedra* and *Gnetum*, both woody members of Gnetales (see Table 1 for a comparative summary of these features). *Ephedra* spp. have Type 5 growth rings (Carlquist 1988), which may be the case with *Aptiana*. Here zones of vessels alternate with vesselless zones of imperforate elements (tracheids in woody Gnetales or fibre-tracheids in *Aptiana*). Both *Aptiana* and *Gnetum gnemon* L. possess vessels with both scalariform and simple perforation plates, usually with an oblique attitude. Scalariform perforation plates in *Gnetum* have few bars, usually 1–7 (Bliss 1921; Maheshwari and Vasil 1961; Muhammad and Sattler 1982). The simple perforations are very similar in both *Aptiana* and *Gnetum*, usually narrow ovate in shape. Ephedroid plates also occur in *Gnetum* (exclusively in *Ephedra*) but no plates of this type are seen in *Aptiana*. Vessel pitting in *Gnetum* is mainly alternate, opposite, or of linear (tracheid-like) arrangement

---

EXPLANATION OF PLATE 2

1–7. *Aptiana radiata*, Stopes, 1912, emend., Lower Cretaceous, Aptian or Albian, Lower Greensand, southern England; The Natural History Museum, London, holotype, NHM V.11517. 1, TLS, vessel with scalariform perforation plate and a uniseriate row of circular pits, NHM V.11517k; ×2920. 2–3, TLS, vessels with alternate circular to oval pitting, NHM V.11517k; ×2920. 4, RLS, simple perforation plate, NHM V.11517b; ×986. 5, RLS, scalariform perforation plate, NHM V.11517h; ×986. 6, RLS, vessel to ray pitting showing possible vestures, NHM V.11517b; ×4640. 7, RLS, unilaterally compound vessel to ray pitting, NHM V.11517p; ×2920.

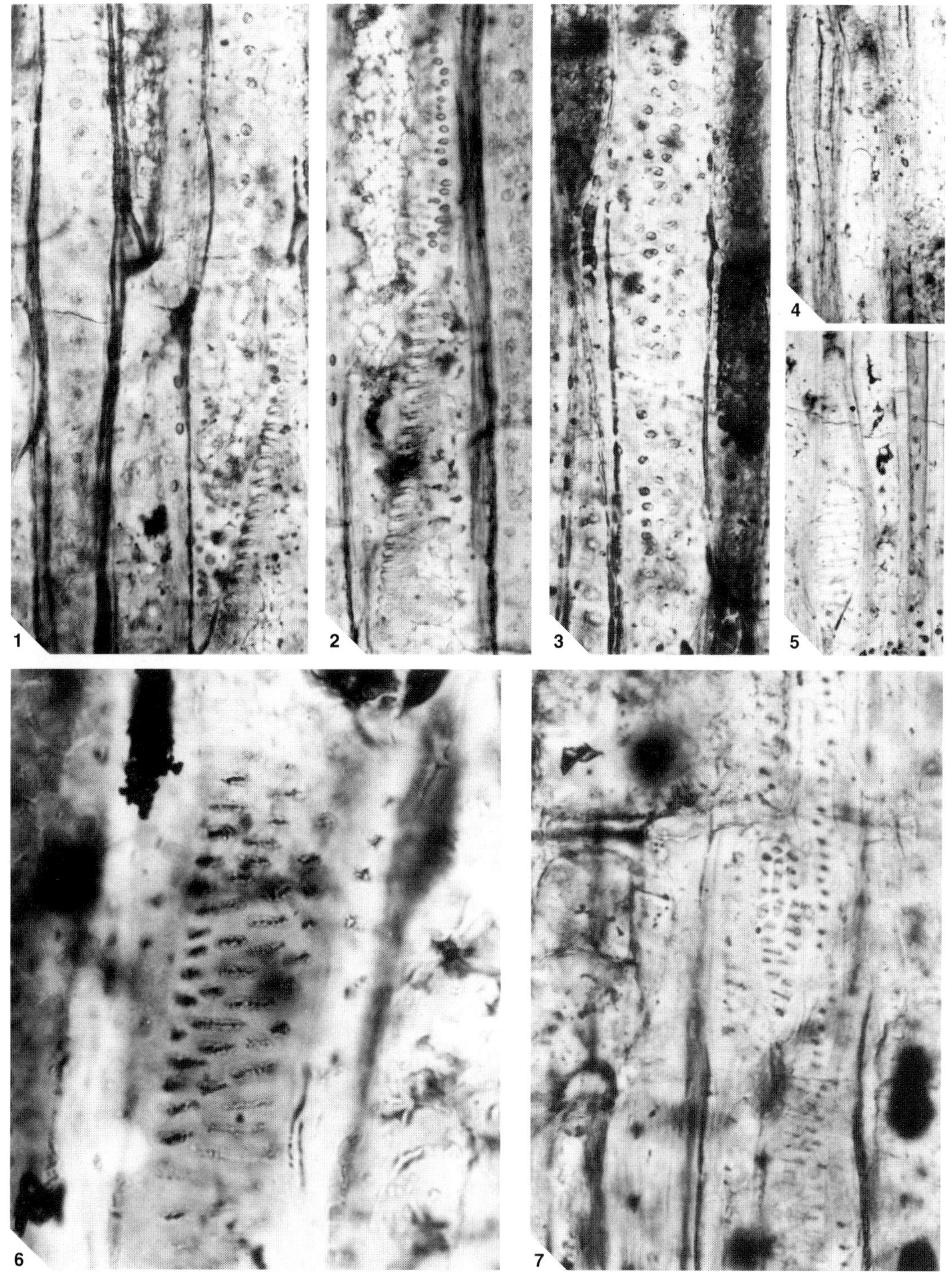

CRAWLEY, *Aptiana*

TABLE 1. Comparison of wood anatomy features of *Aptiana*, dicotyledons and Gnetales.

Feature	*Aptiana radiata*	*Ephedra/Gnetum*	Recent angiosperm
Growth ring/fibre zones	+	+	+
Solitary vessels	+	+	+
Simple and scalariform perforation plates	+	+	+
Vestured pits	+?	+	+
Fibre-tracheids	+	–	+
Heterogeneous rays	+	+	+

(Maheshwari and Vasil 1961, figs 18B, 19–20) rarely scalariform (Muhammad and Sattler 1982, figs 16–17). Individual pits are of variable size (Muhammad and Sattler 1982), bordered, round to oval or elongate with horizontal, slit-like to circular apertures (Chamberlain 1935, fig. 387; Maheshwari and Vasil 1961) and can be vestured (Bierhorst 1960, figs 214, 218; Muhammad and Sattler 1982, figs 67–68). In *Aptiana* the vessel pitting appears to be mainly of alternate type with only some linear arrangement. The pit borders are round, oval or elongate and gash-like, minute to large, and usually with horizontal to oblique (sometimes almost vertical) apertures. Some have structures within the pit aperture that are vesture-like (Pl. 2, fig. 6). This feature is best confirmed by examination of a fractured section under a Scanning Electron Microscope (SEM), as shown by Crawley (1988). *Aptiana*, *Ephedra* and *Gnetum* all possess multiseriate rays showing evidence of aggregation. In *Aptiana* and *Gnetum* the multiseriate rays are spindle-shaped in TLS. Multiseriate rays in *Ephedra* are produced by aggregations and longitudinal division of uniseriate rays (Chamberlain 1935). Aggregation of uniseriate rays with multiseriate rays is seen in *Aptiana* as is some vertical fusion of multiseriate rays (Stopes 1912, text-fig. 5). Stopes (1912) noted 'None of these [multiseriate rays] run all through the stem in transverse section but they dwindle to uniseriate rays, or die out entirely'. Some instances can be quite dramatic (Pl. 1, fig. 4), and this is almost certainly connected with their aggregate nature. This phenomenon possibly occurs in *Ephedra*: Record and Hess (1943, pl. 37, fig. 1) showed a specimen where a multiseriate ray appears and then dwindles to uniseriate. However Carlquist (1989), in an otherwise extensive anatomical study of *Ephedra*, made no mention to it. *Aptiana*, *Ephedra* (four species) and some angiosperm families commonly have uniseriate and multiseriate rays. Ray height is typically under 1 mm for *Aptiana* and *Ephedra* (three species). Ray width is four or more cells wide in *Aptiana* but only 3–4 cells maximum in *Ephedra* spp. with common uniseriate rays (Carlquist 1989). The cellular composition of the rays is heterogeneous in *Aptiana* and *Ephedra* (three species) and homogeneous in *Gnetum* spp. In *Ephedra* the multiseriate ray bodies are composed of procumbent and square cells often externally flanked by sheath cells. There is often only a single upright marginal cell in *Aptiana* and *Ephedra* (Chamberlain 1935; Shavrov 1956; Carlquist 1989). The uniseriate rays are composed entirely of upright cells in *Aptiana* but only partly so in *Ephedra* spp. (Carlquist 1989). In *Aptiana* the widest fused rays sometimes produce a chevron effect when viewed in TS. This feature is characteristic of some species of *Ephedra* (Barefoot and Hankins 1982; Carlquist 1989). In *Ephedra* and *Gnetum* the ground tissue is composed of tracheids. These have a single row of large, circular, bordered pits on their radial surfaces often separated by bars of Sanio (Chamberlain 1935, fig. 347; Bierhorst 1960, figs 206, 208). Pits on the tangential surface have much reduced borders. The pit apertures are round to elongate, horizontal to oblique with crossed pit pairs (Chamberlain 1935, figs 348, 386; Maheshwari and Vasil 1961, fig. 22a, d). Tori are present in *Ephedra* but absent in *Gnetum* (Bliss 1921; Muhammad and Sattler 1982). Tracheid walls are usually thin but are thick in *Ephedra strobilacea* Bunge, 1851 (Shavrov 1956, fig. 2). In *Aptiana* these elements are thick-walled fibre-tracheids with reduced circular to elongate bordered pits on all faces, usually with oblique to vertical slit-like apertures and pit pairs with crossed apertures.

In conclusion, *Aptiana* shares many features with Recent gnetalean wood (see Table 1). However the scalariform perforation plates with many bars, a lack of ephedroid perforations and the presence of fibre-tracheids indicate that it is probably an angiosperm wood of unknown affinity.

Comparison with fossil woods. Four other angiosperm woods are known from Aptian–Albian strata worldwide; *Aplectotremas halisticum* Serlin, 1982, from the Albian of Texas, USA; *Icacinoxylon pittiense* Thayne, Tidwell and Stokes, 1985 from the Aptian–Albian Cedar Mountain Formation, Utah, USA; *Paraphyllanthoxylon idahoense* Spackman, 1948, from Albian strata in Idaho, USA; and *P. utahense* Thayne, Tidwell and Stokes, 1983, also from the Cedar Mountain Formation in Utah. Both *Paraphyllanthoxyllon* and *Aplectrotremas* have vessels in radial multiples, simple perforation plates and septate fibres whereas *Icacinoxylon* has axial parenchyma, opposite vessel-ray pitting and rays more than ten cells wide.

Relatively few Cretaceous angiosperm woods are known. Wheeler *et al.* (1995) counted 110 records of which almost 50 per cent are platanoid/icacinoid/paraphyllanthoid in structure and Wheeler *et al.* (1994) stated that most other Cretaceous woods are a slight variation on these patterns. Other, more recent, specimens have been described from the Upper Cretaceous of Antarctica (Poole and Francis 1999; Poole *et al.* 2000) and Belgium (Meijer 1997). All of these recent specimens have features regarded as primitive in the Baileyan (1924) sense, and are consistent with the variations of pattern noticed by Wheeler.

There is no doubt that angiosperms existed during Aptian–Albian times and probably with gnetaleans (see Crane 1988 for a review of the fossil history of Gnetales). Two possibly gnetalean miospore taxa were recorded from Aptian strata on the Isle of Wight (Atherfield Point, Compton Bay and Redcliff; Atherfield Clay Series, Ferruginous Sands and Sandrock). These remains were rare to extremely rare.

Woods from the Thanet Sand Formation

Chandler (1964) wrote of the Thanet Sand Formation: 'The Thanetian can be dismissed in a few words owing to the paucity of the plant remains yielded.' This 'paucity' is both in numbers and diversity, 11 taxa including six angiosperms, three of which she queried (Chandler 1964; Ward 1978; Collinson 1983). The commonest plant remains are of *in situ* lignified wood with a high pyrite content. These remains are unstable and have not been studied to date. Much rarer silicified woods are also known, are often well preserved, and are discussed here. Seven specimens in The Natural History Museum collections had been collected from Herne Bay, Kent, a classic locality for British Palaeogene sediments. The other four specimens are from Maidstone and the Medway Valley area of Kent, and housed in the collection of Maidstone Museum. However, as it is likely that all the Herne Bay specimens were collected loose from the beach, their original provenance is unclear (D. M. Ward, pers. comm. 1990). Provenance is also in doubt for the Kent specimens, including the holotype material of both *Cantia* and *Hythia*. An attempt to resolve this issue has been made by studying both macro features and the sedimentary fillings of wood borings that are present in some of these specimens. Of critical importance, in my view, is the occurrence of three species at both Herne Bay and in the area around Maidstone: *Anacardioxylon maidstonense* sp. nov, *Cantia arborescens* Stopes, 1915, and *Hythia elgarii* Stopes, 1912. The Maidstone Museum specimens were originally thought to be from the Lower Greensand (Aptian–Albian).

The original provenance of Cantia arborescens *and* Hythia elgarii. Stopes (1915) appeared to be in no doubt that both of these specimens originated from the Hythe Beds of the Lower Greensand. *Cantia arborescens* was found near Ightham, Kent, and *Hythia elgarii* from near Maidstone, Kent. Portions of each were presented to The Natural History Museum by the Committee of the Corporation Museum, Maidstone in 1915. Casey (unpublished BGS report, 1960) concluded the following; '*Cantia arborescens* – There is some sand left in one of the boreholes and this is quite right for the Folkestone Beds of Ightham, Kent, the stated provenance of the type specimen. Conceivably there are other sand formations which would produce a similar matrix but I see no reason to doubt the authenticity of the label. *Hythia elgarii* – There is a little sandy matrix on this specimen, quite in keeping with the Hythe Beds of Maidstone. Some years ago I saw the rest of the type specimen in the Maidstone Museum and decided that it was correctly labelled as from the lower Greensand (Hythe Beds) though at that time I was not aware of the significance of this observation.'

During 1989–1990 additional specimens of *Cantia* (V.57330, V.63150) and *Hythia* (V.794A) were discovered in the collections of The Natural History Museum. They are all from the Thanet Sand

Formation (Palaeocene) of Herne Bay, Kent, and show similar preservation as the types, usually with areas of fungal attack and partial collapse of tissue. A characteristic feature shared by all the specimens is the dark plugging of some vessels in an otherwise very transparent patch of silicification (Pl. 5, figs 1–2; Text-fig. 5C). The Herne Bay specimens are also of similar colour, a whitish or very light grey outer surface and pinkish grey to brownish grey cut surface. Consolidated sandy matrix can be found filling the borings in these specimens, and this has the typical character of sediment from the Thanet Sand Formation (D. W. Ward and J. Cooper, pers. comm. 1990). Re-examination of the type material and the Maidstone Museum specimen of *Anacardioxylan* showed no consolidated rock fillings, only poorly consolidated brick-earth-like material in some borings. In this case I do not feel that Casey's earlier comments are at all conclusive unlike those for *Aptiana* where substantial rock matrix remains. It seems likely that the original provenance of the type material was the Thanet Sand Formation. According to Worssam (1963) there are small outcrops of Thanet Sand Formation in the Maidstone District. Debris has been derived from the formation and is widespread on the Chalk dip-slope surrounding and to the south of these outcrops (see Text-fig. 2). A mechanism for transport of material southwards during the Pleistocene is envisaged by the possibility of non-uniform Pleistocene uplift of the Weald as postulated by Worssam (1973). Certainly the brickearth-like or detrital fillings present in the holotypes at Maidstone Museum would tend to support comparatively recent redeposition. The Maidstone collections also contain other similar silicified dicotyledonous woods from the Medway Gravels, including a *Hythia*-like wood (unnumbered). These specimens had been labelled Lower Greensand but are further examples of wood from the Thanet Sand Formation. The Museum also has a comprehensive collection of bona fide Lower Greensand coniferalean woods which at first glance may appear superficially similar to woods from the Thanet Sand Formation. Both types are extensively bored and iron oxide stained. However, on closer examination Greensand woods show the following differences from those from the Maidstone Thanet Sand Formation: they (1) are always much paler, usually pale cream to white and not pale to dark grey; (2) always have some adherent consolidated matrix; (3) show no signs of external weathering such as the smooth and fluted surfaces seen in *Hythia*; (4) are all coniferalean not angiospermic.

In my view it is important to analyse such general macroscopic features of wood as certain formations yield remains that have recognisably consistent features, such as those from the Thanet Sand Formation and also the Reading Formation. Recognising and accumulating such data can be of particular value when so many specimens of fossil wood are found *ex situ*.

Family ANACARDIACEAE Lindley, 1830

Morphotaxon ANACARDIOXYLON Felix, 1882

Type material. Anacardioxylon spondiaeforme Felix, 1882, Eocene, Caucasus region, USSR.

Anacardioxylon maidstonense sp. nov.

Text-figure 3; Table 2

Derivation of name. After Maidstone Museum, Kent.

Holotype. MM MNE/G/00001; slides MNE/G/00001$1–3.

Paratype. NHM V.794c; slides V.794c$1–3.

Locality and horizon. Reculvers, Herne Bay, Kent (V.794c) and probably Maidstone or Medway area, Kent (MNE/G/00001); Palaeocene Thanet Sand Formation.

Diagnosis. Growth rings, present, indistinct. Vessels diffuse porous, mean density of 23–79 mm^2, solitary vessels (42–52%) and radial groups of 2–4, mean vessel element length 177–318 μm, perforation plates

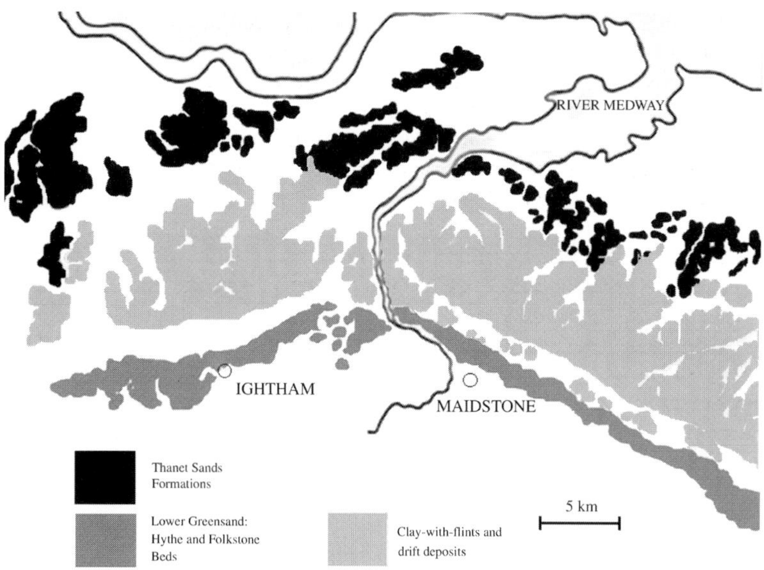

TEXT-FIG. 2. Location of Lower Greensand and Thanet Sand Formation in the Maidstone district (based on British Geological Survey, 1" Series, sheets 271, 272, 287, 288).

exclusively simple, vessel to vessel pitting, alternate, bordered, 5–8 μm in diameter, vessel to parenchyma pitting apparently simple to part-bordered horizontally or vertically elongated. Axial parenchyma paratracheal, vasicentric, sheath one cell thick. Medullary rays 6–9 mm^2, usually 2–3 cells wide or 27 μm, mean height 287 μm or 22 cells, heterogeneous, with a single marginal row of upright cells. Imperforate tracheary elements are libriform fibres. Bark with axial canals.

Description. Two silicified pieces of secondary wood. MNE/G/00001 is a large section of trunk, large branch or root wood whilst V.794c is a branch or root 22 mm in diameter with bark. Growth ring boundaries present but indistinct [**2**], marked by flattened latewood fibres (Text-fig. 3A).

Vessel elements. Diffuse porous [**5**] (Text-fig. 3A, E); solitary vessels 42% (MNE/G/00001) and 52% (V.794c), with radial multiples of 2–4 vessels; simple perforation plates; intervessel pits alternate [**22**], bordered; pit size small [**25**] to medium [**26**], 5–8 μm; vessel-parenchyma pitting with apertures apparently simple to part-bordered, horizontal (gash-like) to almost vertical (palisade) [**32**] (Text-fig. 3D), pit size 6–18 μm. Tangential diameter of vessel lumina: range 33–216 μm, mean 131 μm (MNE/G/00001) [**42**]; range 33–110 μm, mean 79 μm (V.794c) [**41**]. Vessel density 15–34 mm^2, mean 23 (MNE/G/00001) [**48**]; 64–96 mm^2, mean 79 (V.794c) [**49**]. Vessel element length range 220–451 μm, mean 318 μm (MNE/G/00001); range 77–275 μm, mean 177 μm (V.794c) [**52**].

Imperforate tracheary elements. Libriform fibres [**66**], no pitting seen (poorly preserved).

Axial parenchyma. Paratracheal, vasicentric [**79**] as a sheath surrounding vessels, one cell thick (Text-fig. 3F).

Ray parenchyma. Ray width 1–4 cells [**97**]; height range 165–495 μm (12–38 cells), mean 287 μm (22 cells) (MNE/G/00001); width range 16–33 μm (1–4 cells), mean 27 μm (three cells) (MNE/G/00001). Composition is heterogeneous; body ray cells procumbent with one row of upright marginal cells [**106**] (Text-fig. 3C, G). Density is 4–7 rays mm^2, mean 6 (MNE/G/00001), 6–13 mm^2, mean 9 (V.794c) [**115**].

Bark. Composed of cubic cells with dark contents; axial canals present up to 110 × 330 μm, with no discernible contents (Text-fig. 3B).

Comparisons. The main features found in this fossil are; vessels in radial groups of 2–3, with simple perforation plates; alternate and medium-sized vessel to vessel pitting with vessel to parenchyma pitting that is simple and horizontally (gash-like) to vertically (palisade-like) extended; axial parenchyma vasicentric; rays commonly 1–3 seriate and heterogeneous with a single row of upright marginal cells. This combination is found in Recent woods of Anacardiaceae, Buseraceae and

TABLE 2. Comparison of *Anacardioxylon maidstonense* with *A. caracoli*.

		Anacardioxylon maidstonense sp.nov. Upper Palaeocene, Kent V.794c, MM MNE/G/0001	*Anacardioxylon caracoli* G. Schönfeld, 1956 Tertiary, Colombia
Growth rings		indistinct	indistinct
Vessels			
Distribution		diffuse porous	diffuse porous
Grouping		solitary 42% and 52%, radial groups 2–4	solitary, radial groups 2–3
Density/mm^2			
	range	15–96	1–2
	mean	23 and 79	–
Tangential diameter μm			
	range	33–216	150–400
	mean	79 and 131	200–300
Element length μm			
	range	77–451	200–500
	mean	177 and 318	–
Perforation plate		simple	simple
Pitting			
	intervascular	alternate, bordered 5–8 μm	alternate, bordered 12–15 μm
	vessel to parenchyma	simple to part-bordered, horizontal (gash-like) to vertical (palisade), 6–18 μm	part-bordered, round, oval and vertical (palisade), 10–50 μm
Axial parenchyma		paratracheal, vasicentric, sheath 1 cell thick	paratracheal, scanty to vasicentric, sheath 1–3 cells thick
Ray parenchyma			
Density/mm			
	range	4–13	–
	mean	6 and 9	8
Ray width μm			
	range	16–33	–
	mean	27	–
Ray width cells			
	range	2–4	1–2
	mean	2–3	–
Multiseriate height μm			
	range	16–495	–
	mean	287	–
Multiseriate height cells			
	range	12–38	4–30
	mean	22	12
Uniseriate height μm		–	not seen
Composition		heterocellular, single row of upright marginal cells	heterocellular, single row of square to upright marginal cells
Imperforate tracheary elements		libriform fibres	libriform fibres
Axial canals		present in bark	–

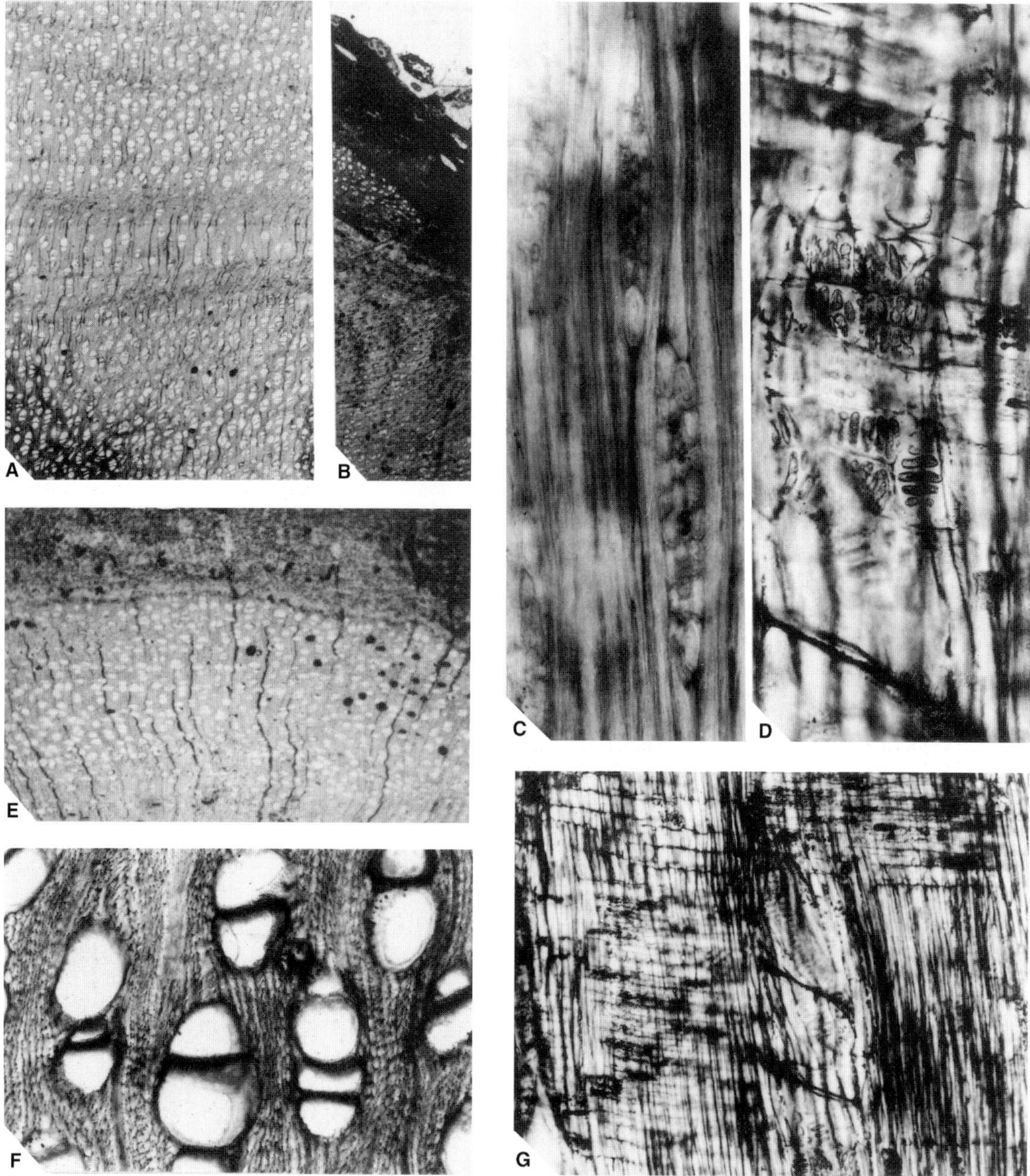

TEXT-FIG. 3. *Anacardioxylon maidstonense* sp. nov. A, C–D, F–G, probably Maidstone or Medway area, Kent, Palaeocene Thanet Sand Formation. B, E, Reculvers, Herne Bay, Kent, Palaeocene Thanet Sand Formation. A, TS, holotype, MM MNE/G/00001$1; ×24. B, TS, showing axial canals in bark, paratype, NHM V.794c$1; ×24. C, TLS, holotype, MMI$2; ×350. D, TLS, vessel to axial parenchyma pitting, holotype, MM MNE/G/00001$2; ×400. E, TS, paratype, NHM V.794c$1; ×30. F, TS, showing paratracheal parenchyma, holotype, MM MNE/G/00001$1; ×200. G, RLS, heterogeneous rays, holotype, MM MNE/G/00001$2; ×250.

Lauraceae. However, V.794c also has axial canals in the bark and both specimens have non-septate libriform fibres that are more typical of Anacardiaceae. The fossil morphotaxa *Anacardioxylon, Boswellioxylon, Burseroxylon, Canarioxylon* and *Ulminium* (*Laurinoxylon*) are similar to the British material. However, because of the absence of septate fibres, idioblasts and radial canals, the wood is assigned to *Anacardioxylon*. It particularly resembles *A. caracoli* G. Schönfeld, 1947. *Anacardioxylon caracoli* is distinct from the Thanet Formation woods in possessing substantially larger vessels, much larger pitting and wider sheaths of vasicentric parenchyma. The material is, therefore, placed in a new species, *A. maidstonense*.

Remarks. Although the specimens described here are quantatively different they are qualitatively similar. This is almost certainly intraspecific variation owing to relative position on the tree. V.794c is a small axis whilst the Maidstone Museum specimen is part of a substantial branch, trunk or root. Secondary wood elements tend to be much smaller in the smaller axes of the same tree or shrub (Barefoot and Hankins 1982). Another anacardiaceous wood is known from the Thanet Sand Formation (see *Edenoxylon aemulum* Brett, 1966 below). Anacardiaceous fruit, seed and pollen remains have not been recorded below the Woolwich and Reading Formation (Chandler 1964).

The specimens are the first example in this study of the same species occurring at both Herne Bay and in the Maidstone area, the others being *Cantia arborescens* Stopes, 1915 and *Hythia elgarii* (Stopes) comb. nov. The Maidstone Museum specimen was originally thought to be from the Lower Greensand (Aptian) (see section on *Cantia arborescens* for discussion of original provenance).

Morphotaxon EDENOXYLON Kruse, 1954, emend.

Type species. *Edenoxylon parviareolatum* Kruse, 1954, Eocene, Eden Valley, Wyoming.

Emended diagnosis. Vessels diffuse to semi-ring porous. Rays narrow, 1–3 seriate. Imperforate tracheary elements, libriform fibres, septate or non-septate.

Edenoxylon aemulum Brett, 1966, emend.

Plate 3; Table 3.

1966 *Edenoxylon aemulum* Brett, p. 360, Pl. 61, figs 3–7.

Emended diagnosis. Growth rings absent to distinct. Diffuse to semi-ring porous. Rays 8–15 mm^2, 6–16 cells high. Imperforate tracheary elements are libriform fibres.

Material. NHM V.63151; slides $1–3.

Locality and horizon. Herne Bay, Kent; Palaeocene Thanet Sand Formation.

Description. Growth rings are absent to distinct [**1–2**], marked by larger, denser vessels in early-wood (Pl. 3, fig. 1).

Vessel elements. Semi-ring porous [**4**] to diffuse [**5**] (Pl. 3, fig. 1), solitary and with radial multiples of 2–4, mostly two (Pl. 3, fig. 3); perforation plates exclusively simple [**13**]; vessel to vessel pitting alternate, bordered, small, 4–6 μm in diameter [**25**] (Pl. 3, fig. 2); pits to parenchyma apparently simple to part-bordered, round, angular or horizontal (gash-like) in shape, up to 20 μm across [**32**] (Pl. 3, fig. 4); vessels very small to moderately small. Tangential diameter of vessel lumina: range 33–121 μm, mean 77 μm [**41**]. Density is 30–66 per mm^2 [**49**] (inner to outer ring); element length, short, range 275–467 μm, mean 313 μm [**52**]; thin-walled tyloses present [**56**].

Imperforate tracheary elements. Libriform fibres [**66**] thin-walled [**68**], length and pitting not observed.

Axial parenchyma. Paratracheal, vasicentric [**79**], sheaths one cell thick.

Ray parenchyma (*excluding those containing radial canals*). Rays usually two cells wide (52%) [**97**]; uniseriate rays common (22%), very low to low. Multiseriate ray height: range 132–297 μm, (6–16 cells), mean 200 μm (ten cells);

uniseriate ray height: range 55–165 μm (2–9 cells), mean 99 μm (four cells). Ray widths fine, 16–39 μm (1–3 cells). Composition is heterogeneous; multiseriate rays composed of procumbent cells, usually with a single marginal row of upright cells [**106**] (Pl. 3, fig. 6); uniseriate rays composed of procumbent and upright cells. Density is 9–15 mm^2 [**116**].

Storied structure. Approaching en échelon arrangement [**122**].

Intercellular canals. Normal radial ducts present within multiseriate rays [**130**] (Pl. 3, fig. 5); ray height 330–770 μm (20–40 cells), width 100–275 μm (7–13 cells). Ducts usually one per ray, oval; height 88–352 μm, width 55–162 μm, surrounded by 3–5 epithelial layers.

Discussion. As a result of its better preservation the new specimen has revealed further anatomical details of this taxon. *Edenoxylon* Kruse, 1954, as emended by Brett (1966), was erected for fossil anacardiaceous woods not comparable with any living genus but possessing features characteristic of the tribes Spondieae and Rhoideae. These features were described by Brett as the presence of vessel multiples; scanty paratracheal parenchyma; radial ducts; septate fibres; minute intervascular pitting; and small areas of opposite intervascular pitting. Re-examination of the holotype specimen did not reveal the presence of septate fibres as observed by Brett. Most of the fibres are full of gum. In many cases these contents are

TABLE 3. *Edenoxylon aemulum*: comparison of mature wood specimens.

		V.44297	V.62151
Growth rings		present	present
Vessels			
Distribution		diffuse to semi-ring porous	diffuse to semi-ring porous
Density/mm^2			
	mean	44	30
Tangential diameter μm			
	range	30–140	33–121
	mean	95	77
Element length μm			
	range	270–440	275–467
	mean	345	313
Perforation plate		simple	simple
Pitting			
	intervascular	alternate, bordered 5μm	alternate, bordered 4–6 μm
	vessel to parenchyma	simple, round to gash-like	simple, round to gash-like
Axial parenchyma		paratracheal, scanty	paratracheal, scanty
Ray parenchyma			
Density/mm			
	range	6–12	9–15
Ray width cells			
	range	1–2	1–3
	mean	2	2
Multiseriate height μm			
	range	110–280	132–297
	mean	183	200
Multiseriate height cells			
	range	5–14	6–16
	mean	10	10
Composition		heterocellular	heterocellular
Imperforate tracheary elements		thin walled libriform fibres	thin walled libriform fibres
Radial canals		present, up to 250 μm diameter and up to 1000 μm high	present, up to 275 μm diameter and up to 770 μm high

separated transversely by spaces that are not very deep and may give the impression of a septum keeping the gum plugs apart. Brett also observed that some broader rings have a definite late-wood with smaller and fewer vessels but considered the woods to be diffuse porous. I interpret this as a diffuse to semi-ring porous arrangement of the vessels. Diffuse to semi-ring porous vessel distribution and non-septate fibres are features of the new specimen (V.63151) and hence are included in the emended species diagnosis. This does not affect Brett's intended use of the genus as the features of the fossil still show that it is a member of Anacardiaceae (the anatomically similar woods of the Burseraceae have exclusively septate fibres).

Remarks. This is undoubtedly another specimen of *Edenoxylon aemulum* Brett, 1966. Quantitative and qualitative features of the new specimen are compared to the holotype in Table 3. The minor differences, e.g. some three-seriate rays and possible vasicentric parenchyma, are acceptable intraspecific variations probably representing different areas of the tree (branch-wood, root-wood, trunk-wood).

Typical features of the Anacardiaceae are vasicentric parenchyma, rays that are short and fine, and radial canals. Genera with semi-ring porous to ring porous vessel distribution and non-septate fibres occur only in the tribes Spondieae and Rhoideae (Heimsch 1942; Kryn 1952). The fossil differs, however, from all Recent taxa within these tribes. Ring porous members of the Rhoideae have spiral thickening in the vessels whilst those of Spondieae have septate fibres. However the anatomical characters taken as a whole indicate that this wood represents an extinct member or ancestor of the Spondieae or Rhoideae. Several other fossil anacardiaceous woods possibly representing Spondieae or Rhoideae have been described (Kruse 1954; Prakash and Tripathi 1969; Dupéron 1973; Schilkina 1973; Manchester 1977; Wheeler *et al.* 1978; Ghosh and Roy 1979; Crawley 1989). Besides *Edenoxylon aemulum* only two other species show any degree of ring porosity: *Pistacioxylon multicoides* Dupéron, 1973 from the Oligocene of south-west France; and *Rhus crystallifera* Wheeler, Scott and Barghoorn, 1978, from the Eocene of Yellowstone National Park, USA. Both specimens of *Edenoxylon* show abundant borings, some of which provide evidence of insect attack. The characteristic polygonal faecal pellets present have been identified as being from dry-wood termites of Kalotermitidae (Dr W. Sands, pers. comm. 1990)

Family AQUIFOLIACEAE Bartling, 1830

Morphotaxon ILICOXYLON E. Schönfeld, 1956

Type species. *Ilicoxylon ilicoides* Schönfeld, 1956, Palaeogene, Iceland.

Ilicoxylon? *prestwichii* sp. nov.

Plate 4

Derivation of name. After the collector of the first specimen, Prof. J. Prestwich. His pioneering work on British Palaeogene stratigraphy includes the recognition and naming of the Thanet Sand Formation (Prestwich, 1852).

Holotype. NHM. V.794B; slides $V.794B1–3.

Paratype. NHM. V.27923; slides V.27923$1–9.

EXPLANATION OF PLATE 3

1–6. *Edenoxylon aemulum* Brett, 1966, emend. Herne Bay, Kent, Palaeocene Thanet Sand Formation, NHM V.63151. 1, TS, NHM V.63151$3; ×12. 2, RLS, intervascular pitting. NHM V.63151$2; ×400. 3, TS, NHM V.63151$3; ×80. 4, RLS, vessel to ray pitting, NHM V.63151$2; ×400. 5, TLS, NHM V.63151$1; ×80. 6, RLS, NHM V.63151$2; ×80.

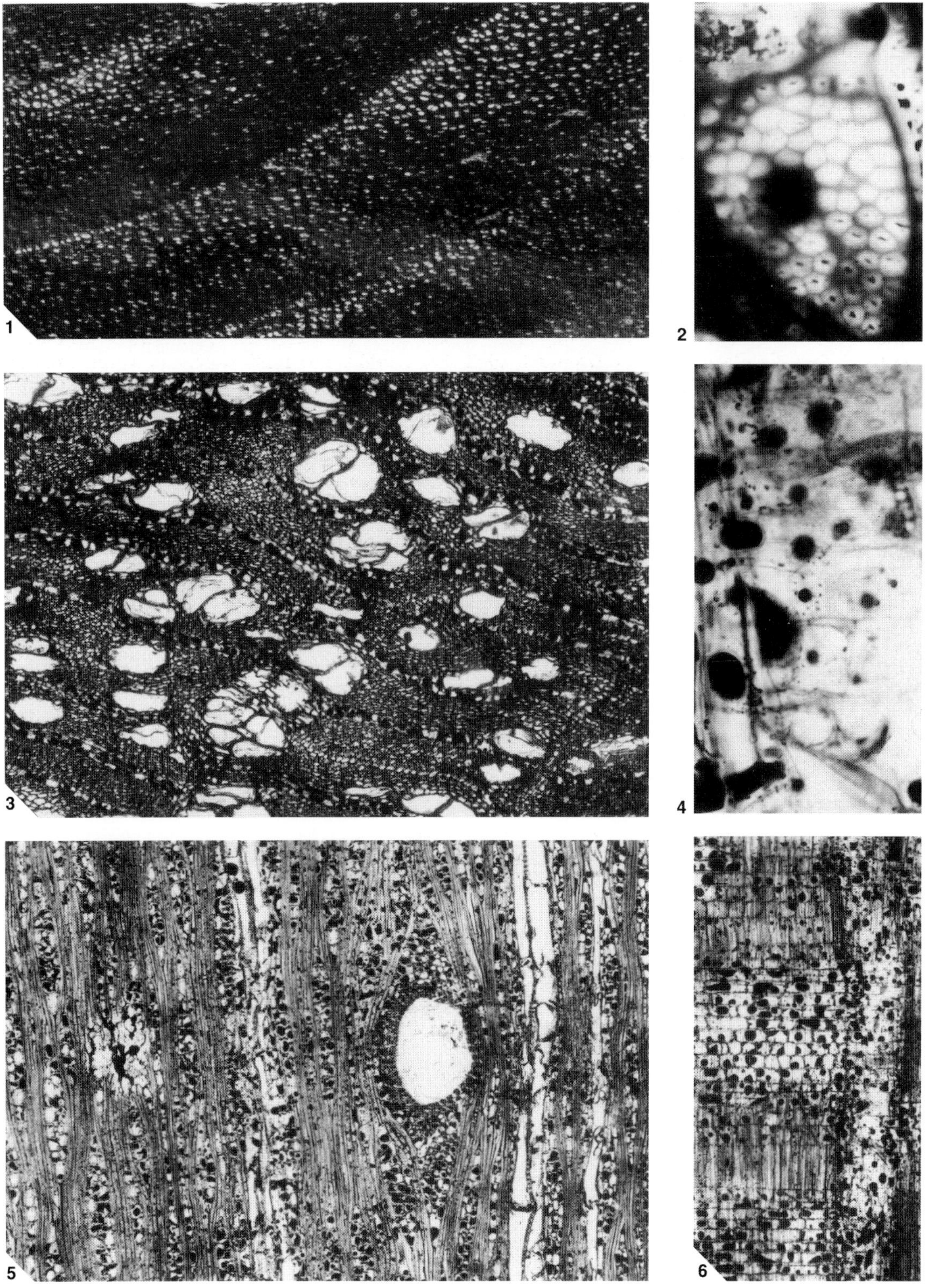

CRAWLEY, *Edenoxylon*

Locality and horizon. Reculvers (V.794B), Herne Bay (V.27923), Kent; Palaeocene Thanet Sand Formation.

Diagnosis. Vessel elements diffuse porous, at least 50 mm^2; solitary and in radial multiples of up to eight, tangential diameter up to 36 μm, perforation plates exclusively scalariform with 24–48 bars, usually 32, vessel to vessel pitting transitional, pitting to parenchyma similar to intervascular. Axial parenchyma apotracheal, diffuse to diffuse-in-aggregate. Rays 12–17 mm^2; of two sizes, with uniseriate rays common (40%), mean multiseriate ray heights 1059–1166 μm or 18–25 cells, mean uniseriate ray heights 555–720 μm or 8–13 cells, 1–8 cells wide, mean multiseriate ray widths 120–140 μm or 4–5 cells, mean uniseriate ray widths 26–29 μm, heterogeneous, multiseriate ray margins of 2–7 square to upright cells. Fibre-tracheids with distinctly bordered pits.

Description. No growth rings observed [**2**].

Vessel elements. Diffuse porous [**5**], solitary and in radial groups of up to eight [**10**] (Pl. 4, figs 1–3). Perforation plates exclusively scalariform [**14**] with 24–48 bars, usually 32 [**18**] (Pl. 4, figs 4–5). Intervascular pitting transitional [**20, 21**] (Pl. 4, fig. 6); pitting to parenchyma similar to intervascular [**30**]. Vessels are rectangular to subcircular in section, very small; tangential diameter up to 36 μm [**40**]; length not observed. Density estimated to be at least 50 per mm^2 [**49**] (overlap in size with fibre-tracheids precludes a complete assessment).

Imperforate tracheary elements. Fibre-tracheids with distinctly bordered pits [**62**]; length not observed; width 12–26 μm.

Axial parenchyma. Apotracheal, diffuse [**76**] to diffuse-in-aggregate [**77**] as short uniseriate chains between rays.

Ray parenchyma. Rays fine to wide, 1–8 cells [**98**], moderately high [**102**]; multiseriate ray height: range 405–1710 μm (6–35 cells), means 1059 and 1166 μm (18 and 25 cells); uniseriate ray height: range 450–990 μm (6–15 cells), means 555 and 720 μm (8 and 13 cells). Rays of two sizes [**103**] with uniseriate rays common (40%); multiseriate rays 70–265 μm, means 120 and 140 μm (four and five cells); uniseriate rays 20–35 μm, means 26 and 29 μm. Composition is heterogeneous; multiseriate rays composed of procumbent and upright cells and 2–7 (usually five) [**108**] square to upright marginal cells; uniseriate rays composed of square to upright cells; occasional sheath cells present [**110**] (Pl. 4, fig. 2). Density is 12–17 mm^2 [**116**].

Remarks. The secondary wood of many dicotyledonous families possesses the following combination of unspecialized features that are present in these fossils: transitional pitting; scalariform perforation plates with many bars; dimorphous, high and markedly heterogeneous rays; diffuse parenchyma; fibre-tracheids. However the distinct radial grouping of the vessels is typical of Aquifoliaceae. Baas (1973, 1975) has given a detailed account of the wood anatomy of Aquifoliaceae from which it appears that the fossils are most similar to *Ilex*. Four species of fossil aquifoliaceous wood have been described: *Ilex* cf. *aquifolium* of Hofmann, 1934; *Ilicoxylon austriacum* Selmeier, 1970; *I. ilicoides* Schönfeld, 1956; and *I. theresiae* Greguss, 1969. *Riboidoxylon cretacea* Page, 1970 from the Upper Cretaceous of California also resembles the wood of *Ilex*. All are distinct from the Herne Bay woods. *Ilex* cf. *aquifolium* and *Ilicoxylon ilicoides* have multiseriate rays that are usually only 2–3 cells wide; *I. theresiae* has scalariform perforation plates with only 10–15 bars; *I. austriacum* has 10–33 bars and scalariform pitting; and *Riboidoxylon cretacea* 20–30 bars and extremely tall rays (up to 9 mm). The erection of a new species of *Ilicoxylon*, *I. prestwichii* sp. nov. is, therefore, justified.

Aquifoliaceous remains are not common in the British lower Tertiary but pollen and macrofossil remains assigned to *Ilex* are known from Eocene and Oligocene formations in south-east England (Boulter and Hubbard 1982; Chandler 1964).

EXPLANATION OF PLATE 4

1–6. *Ilicoxylon*? *prestwichii* sp. nov., Reculvers, Herne Bay, Kent; Palaeocene Thanet Sand Formation, holotype, NHM V.794B. 1, TS, NHM V.794B$1; ×240. 2, TLS, multiseriate rays showing sheath cells, NHM V.794B$2; ×240. 3, TS, NHM V.794B$1; ×160. 4–5, RLS, scalariform perforation plates, NHM V.794B$3; ×350. 6, RLS, intervascular pitting, NHM V.794B$3; ×400.

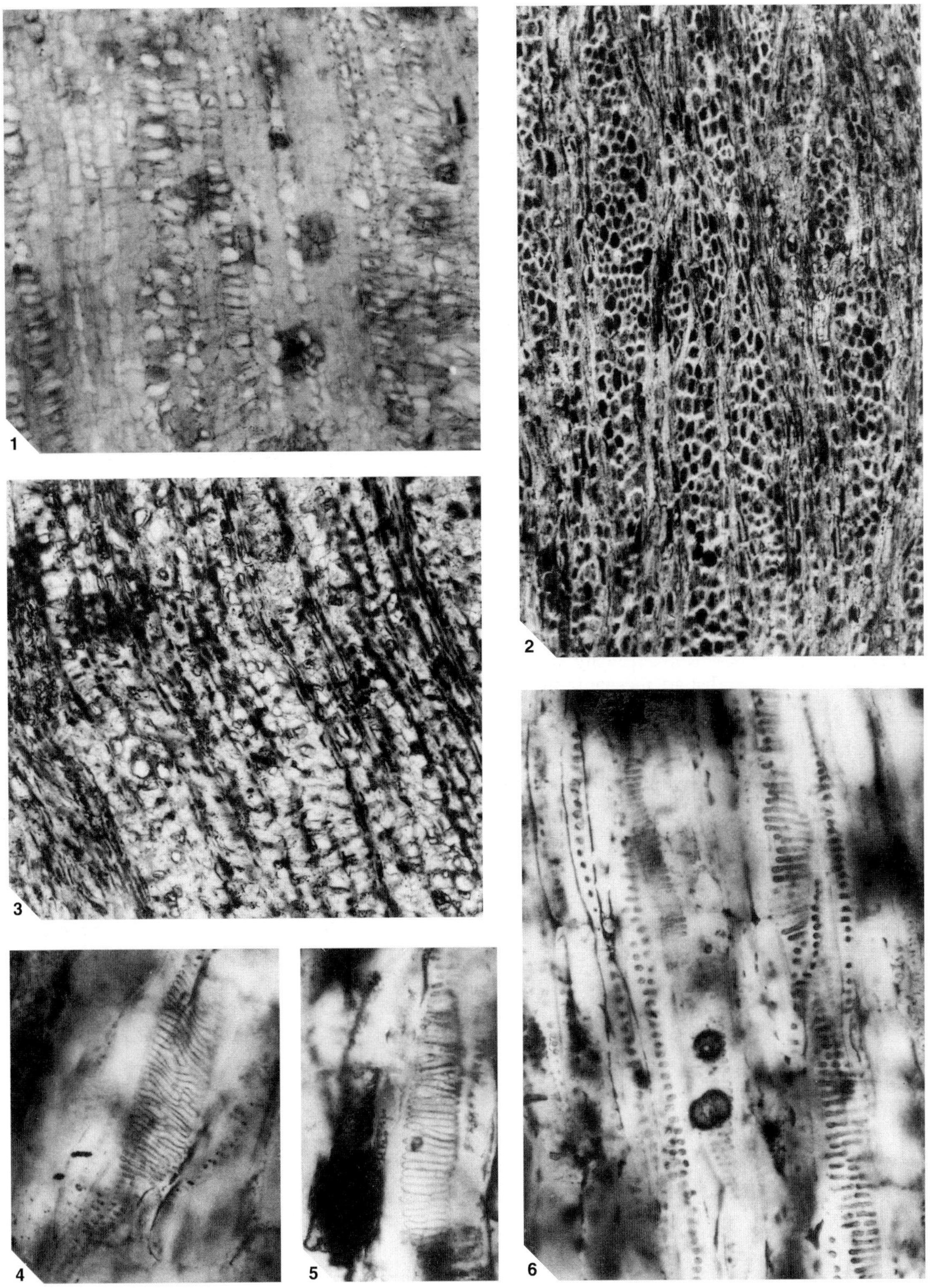

CRAWLEY, *Ilicoxylon*?

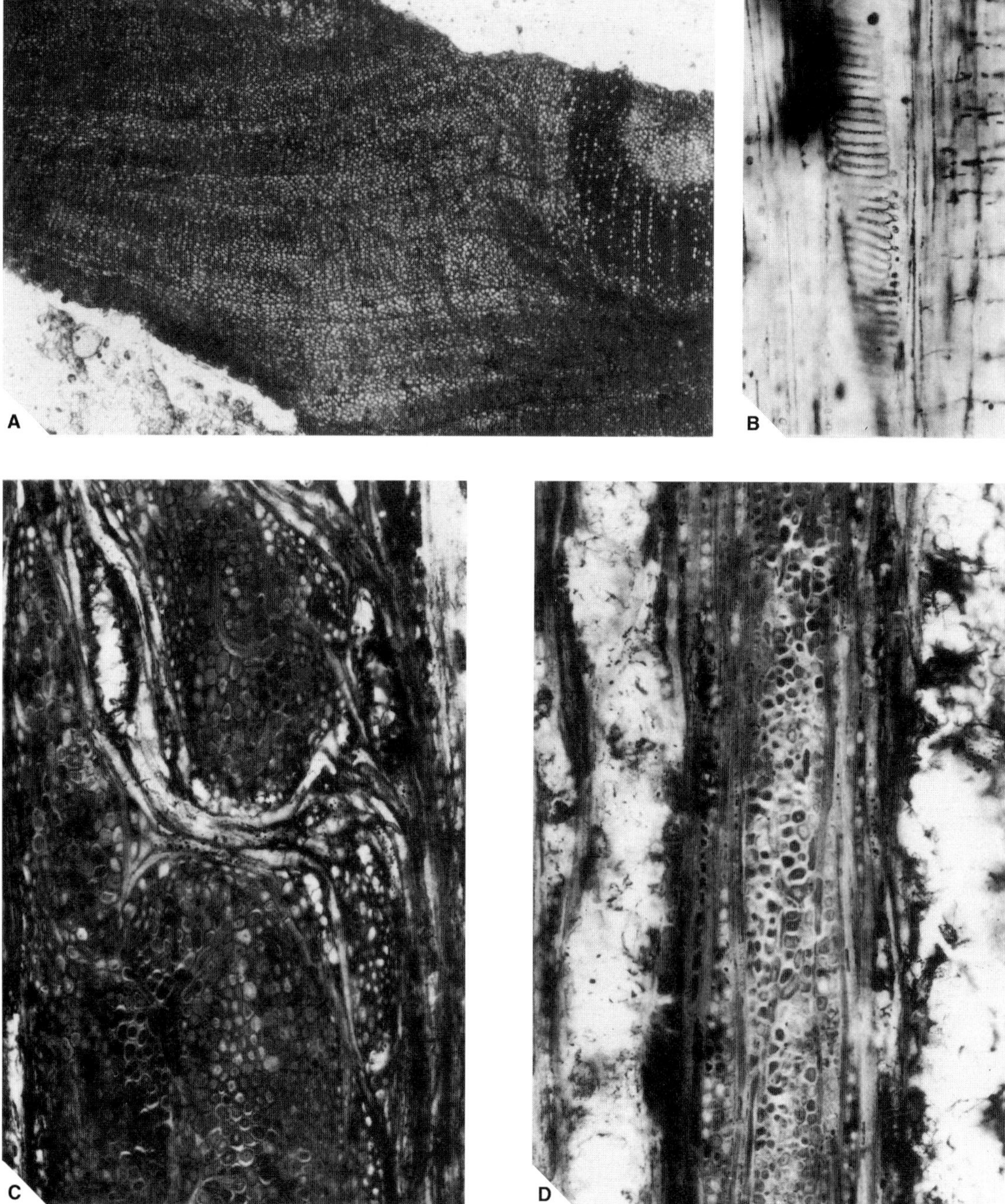

TEXT-FIG. 4. *Cantia arborescens*, Herne Bay, Kent, Palaeocene, between Morrisi and Corbula Beds, Thanet Sand Formation. A, TS, paratype, NHM V.63150$1, ×10. B, RLS, scalariform perforation plate, paratype, NHM V.57330$2; ×300. C, TLS, aggregate ray, paratype, NHM V.63150$2; ×64. D, TLS, aggregate ray with flanking uniseriate rays, paratype, NHM V.63150 $2; ×160.

Family BETULACEAE Gray, 1821

Morphotaxon CANTIA Stopes, 1915

Type species. Cantia arborescens Stopes, 1915, near Ightham, Kent; derived from the Palaeocene Thanet Sand Formation (see introduction to Thanet Sand Formation for a discussion of the provenance of *Cantia* and *Hythia*).

Diagnosis. Growth rings present and indistinct or absent. Diffuse porous, with solitary vessels and some radial and tangential groups. Pitting scalariform to alternate. Perforation plates usually scalariform; some simple plates may be present. Parenchyma diffuse. Rays uniseriate to triseriate, aggregate rays present or absent. Imperforate tracheary elements, fibre-tracheids.

Cantia arborescens Stopes, 1915, emend.

Plate 5; Text-figure 4; Tables 3–4

1915 *Cantia arborescens* Stopes, p. 260, pls 26–28; text-figs 76–78.
1924 *Cantia arborescens* Stopes; Scott, p. 53.
1931 *Cantia arborescens* Stopes; Edwards, p. 27.
1971 *Cantia arborescens* Stopes; Pant and Kidwai, p. 252.

Holotype. NHM V.13231; slides V.13231a–i and MM MNE/G/00003, Maidstone Museum and Art Gallery.

Paratypes. NHM V.57330; slides V.57330$1–4; V.63150; slides V.63150$1–12.

Other material. MM MNE/G/00002.

Locality and horizon. As holotype (MM MNE/G/00002) and Herne Bay, Kent (V.57330, V.63150); Palaeocene, between Morrisi and Corbula Beds (V.57330), Thanet Sand Formation.

Diagnosis. Growth rings present, indistinct. Mean vessel density 125–160 mm^2. Pitting is mainly transitional to opposite, sometimes scalariform. Perforation plates usually scalariform with 13–34 bars. Rays exclusively uniseriate, mean height 377–393 (11–19 cells). Aggregate rays composed of multiseriate units present or absent.

Description. Growth rings distinct to indistinct [**1–2**]; where present the ring boundary is marked by a row of flattened fibres and fewer vessels.

Vessel elements. Vessels diffuse porous [**5**] (Text-fig. 4A), often with radial alignment [**7**]; mainly solitary and in radial, oblique and tangential groups of 2–6 (Pl. 5, figs 1–2); angular in section [**12**]. Perforation plates, some simple [**13**] but mainly scalariform [**14**] with 13–33 bars [**16–17**] (Text-fig. 4B). Vessel to vessel pitting, bordered, mainly transitional but areas of scalariform [**20**], opposite [**21**] and alternate [**22**] also occur; pit apertures 3–28 μm across [**28**]; pits to parenchyma similar to intervascular [**30**] (Pl. 5, figs 3–7). Tangential diameter of vessel lumina, small to very small, range 22–66 μm, means 50 and 55 μm [**41**]. Density 100–175 mm^2, means 133, 160 [**50**]. Vessel element length range 385–990 μm, means 563 and 625 μm [**53**]; tyloses and gum deposits present locally [**56, 58**].

Imperforate tracheary elements. Fibre-tracheids [**62**]; length 530–880 μm, tangential diameter 11–20 μm.

Axial parenchyma. Apotracheal, diffuse [**76**].

Ray parenchyma. Ordinary rays uniseriate [**96**] (Text-fig. 4D) with rare biseriate portions; aggregate rays present [**101**]; ray height range 71–605 μm (3–28 cells), means 340 and 370 μm (15 and 17 cells). Ray width range 24–28 μm (1 cell). Composition is homocellular to weakly heterogeneous [**109**]; aggregate rays composed mainly of multiseriate ray units that are regular to irregular in shape, interspersed with parenchyma strands (Text-fig. 4C–D). Aggregate ray width range 225–450 μm. Density 10–14 per tangential mm [**115–116**]; gum deposits common.

Bark. A zone 5 mm broad consisting of sclerotic xylem tissue with rays up to three cells wide.

Comparisons. This species is undoubtedly a member of the Betulaceae. Woods of this family are typified by the presence of aggregate rays, perforation plates that are sclariform to simple, vascular tracheids and/or fibre-tracheids, heterogeneous to homogeneous rays and apotracheal parenchyma. All of these are present

TABLE 4. Comparison of *Cantia arborescens* with the wood of extant Betulaceae. MVD, mean vessel density; MVDi, mean vessel diameter; MEL, mean element length; IVPT, intervascular pitting transitional; PPS, perforation plates scalariform; PPSi, perforation plates simple; ST, spiral thickenings in vessels; TR, tracheids; FT, fibre-tracheids; AR, aggregate rays; HR, heterogeneous rays; HoR, homocellular rays; RS, ray seriation; +, present; (+), present but not predominant; – not present; information on Recent genera after Hall (1952).

	Cantia	*Betula*	*Alnus*	*Ostryopsis*	*Carpinus*	*Ostrya*
MVD mm^2	133–160	24–211	35–150	175	30–124	28–138
MVDi μm	50–55	30–90	25–65	20–40	25–80	13–95
MEL μm	563–625	430–775	376–878	351	505–949	541–913
IVPT	+	–	(+)	–	–	–
PPS	+	–	+	–	(+)	(+)
PPSi	(+)	–	–	–	(+)	+
ST	–	–	–	+	+	+
TR	+	–	+		–	–
FT	–	–	+	+	+	+
AR	–	(+)	+	+	+	–
HR	(+)	(+)	–	+	–	–
HoR	(+)	–	+	(+)	–	–
RS	1	1–3–(5)	1–(2–3)	1–2	1–3–(5)	1–3–(5)

in the fossil. *Cantia* is compared to all extant genera of Betulaceae in Table 4. Although there is not an exact match, the fossil is most similar to woods of the tribe Betuleae, particularly *Alnus*. Marie Stopes and L. A. Boodle (Kew) had also made tentative comparisons with *Alnus* and Betulaceae (as well as other families). *Alnus* is considered the least specialized of all the Recent genera (Hall 1952). Some species possess transitional vessel pitting, although this is not predominant. Scalariform pitting is not recorded in Recent Betulaceae but is present in the Herne Bay fossils. For these reasons the fossil cannot be regarded as an analogue of any extant genus and is regarded as a probable extinct member of the tribe Betuleae.

Other fossil wood of Betuleae is restricted to the Tertiary of Europe and the USA (Müller-Stoll and Mädel 1959; Prakash 1968; Greguss 1969; Petrescu and Nutu 1969; Wheeler *et al.* 1977; Scott and Wheeler 1982; Blokhina and Snezhova 1999). They are cited as *Alnus* or *Betula*, or referred to the morphotaxa *Alnoxylon* Felix, 1884 or *Betuloxylon* Kaiser, 1880. They differ from *Cantia* in exactly the same features as Recent Betuleae, i.e. in their vessel pitting. Steward and Holltum (1924) mentioned a possible betulaceous wood from the Palaeocene of Mull, Scotland, but this was never described.

Remarks. The current specimens are very similar to the holotype (see Table 5) and differ only in some variation between the respective intervascular pitting and the presence of aggregate rays in one (V.63150). In the holotype the pitting is mainly opposite, with transitional and alternate also present, whilst the Herne Bay woods have mainly transitional pitting with scalariform, opposite and alternate. This type of variation is seen in Recent *Alnus* where transitional (near the pith) to opposite and alternate pitting (some distance from the pith, formed by older cambium) can occur within a single tree (Hall 1952). Betulaceous leaves

EXPLANATION OF PLATE 5

1–7. *Cantia arborescens* Stopes, 1915, emend.: 1, 3–4, near Ightham, Kent, derived from the Palaeocene Thanet Sand Formation and 2, 5–7 from Herne Bay, Kent; uppermost Palaeocene/lowermost Eocene, between Morrisi and Corbula Beds, Thanet Sand Formation. 1, TS, characteristic dark plugging of vessels, holotype, NHM 13231a; ×100. 2, TS, similar dark plutting, paratype, NHM V.63150; ×100. 3, RLS, mainly opposite vessel to ray pitting, holotype, NHM V.13231d; ×1856. 4, RLS, transitional vessel to ray pitting, holotype, NHM V.13231d; ×1856. 5, RLS, mainly opposite vessel to ray pitting, paratype, NHM V.57330$2; ×1856. 6, RLS, mainly transitional vessel to ray pitting, paratype, NHM V.57330$2; ×1856. 7, RLS, scalariform vessel to ray pitting, paratype, NHM V.57330$2; ×1856.

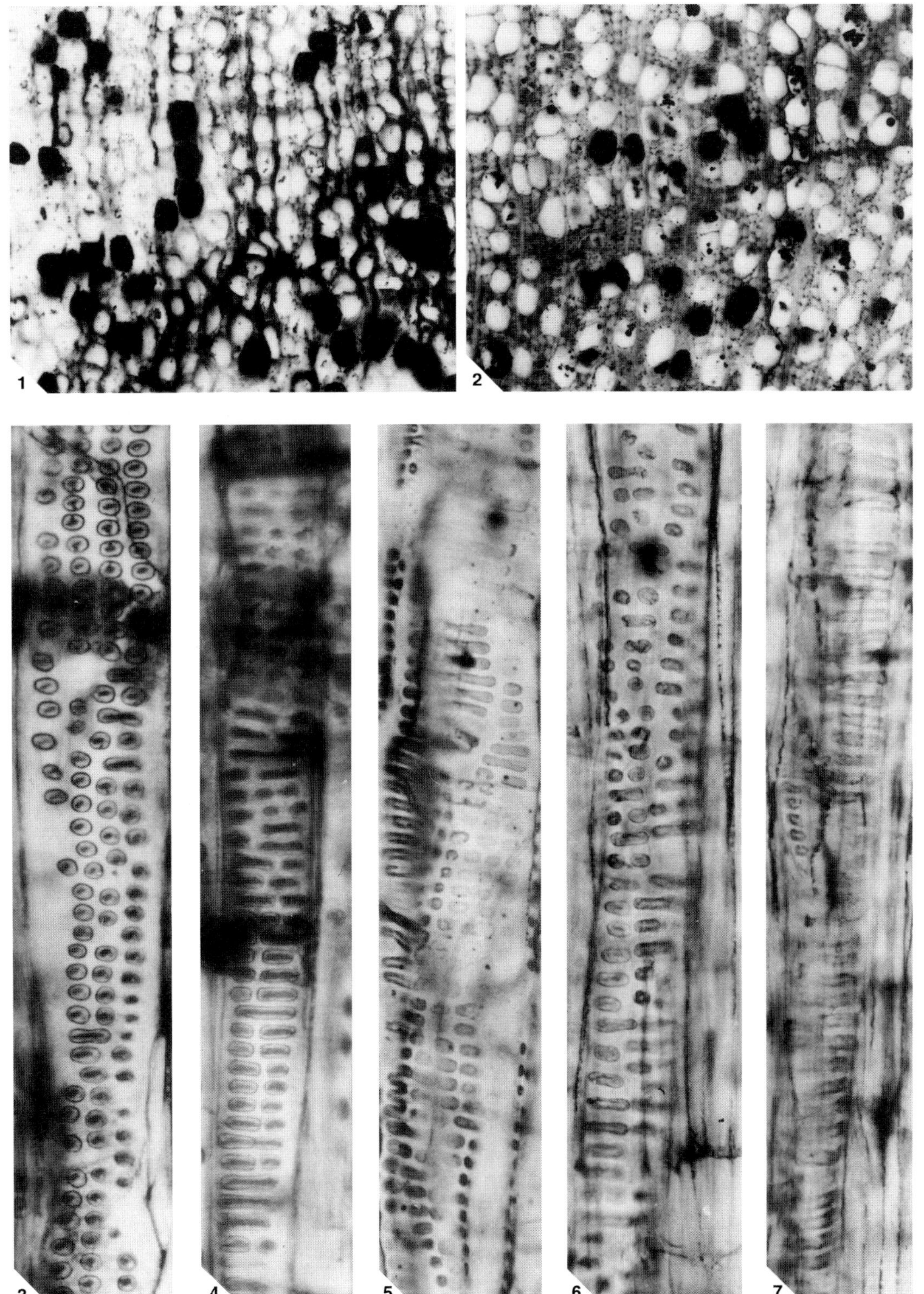

CRAWLEY, *Cantia*

TABLE 5. *Cantia arborescens*: comparison of the holotype with two further specimens from the Thanet Sand Formation.

	V.57330, V.63150	V.13231, holotype
Growth rings	present	present
Vessels		
Distribution		
Grouping	mostly solitary and in radial to tangential groups 2–4	mostly solitary and in radial to tangential groups of 2–3
Density/mm^2 mean	133 and 160	125
Tangential diameter μm mean	50 and 55	42
Perforation plate	scalariform with 13–33 bars	scalariform with 17–34 bars
Pitting		
intervascular	bordered, mainly transitional, also scalariform, opposite and alternate	bordered, mainly opposite, some transitional and alternate
vessel to parenchyma	similar	similar
Axial parenchyma	apotracheal, diffuse aggregate rays	probably apotracheal and diffuse
Ray parenchyma	present in V.63150; normal rays exclusively uniseriate and 3 seriate in bark	exclusively uniseriate
Density/mm mean	12 and 13	17
Ray height μm mean	390 and 393	377
Ray height cells mean	11 and 12	19
Composition	homocellular to heterocellular	homocellular
Imperforate tracheary elements	fibre-tracheids	fibre-tracheids

and infructescences are known from the Palaeogene of Britain as follows: *Carpinus davisii* Chandler, 1961, *Craspedodromophyllum acutum* Crane, 1981 and *Palaeocarpinus laciniata* Crane, 1981 (all probably Coryleae) from the Woolwich and Reading Formation; and *Alnus poolensis* Chandler, 1963, from the Middle Eocene Bournemouth Freshwater Beds. Pollen grains of *Alnus*, *Betula* and *Carpinus* have been recorded from the Palaeogene of south-east England (Chandler 1964; Boulter and Hubbard 1982; Boulter and Kvaček 1989).

The Betuleae are recognized as a distinct group from the Late Cretaceous onwards (see Crane and Stockey 1987 for a review of the literature). Most of the records are indicative of *Alnus* or *Betula*-like plants but at least one group of infructescences may indicate an extinct genus of Betuleae in the Palaeocene of North America (Crane and Stockey 1987).

The Thanet Sands woods show abundant borings, some of which provide evidence of insect attack. The characteristic polygonal faecal pellets present have been identified as being from dry-wood termites of Kalotermitidae (Dr W. Sands, pers. comm. 1990)

Family EUPHORBIACEAE Jussieu, 1789

Morphotaxon EUPHORBIOXYLON Felix, 1887

EXPLANATION OF PLATE 6

1–6. *Euphorbioxylon hernense* sp. nov., Reculvers, Herne Bay, Kent; Palaeocene Thanet Sand Formation, holotype, NHM 794D. 1, TS, showing axial canals in tangential series, NHM V.794D$1; ×24. 2, TLS, multiseriate and uniseriate rays, NHM V.794D$2; ×240. 3, TS, NHM V.794D; ×300. 4, RLS, heterogeneous rays, NHM V.794D$3; ×240. 5, TS, axial canals, NHM V.794D$1; ×320. 6, RLS, showing vessel to vessel pitting and a simple perforation plate, NHM V.794D$3; ×500.

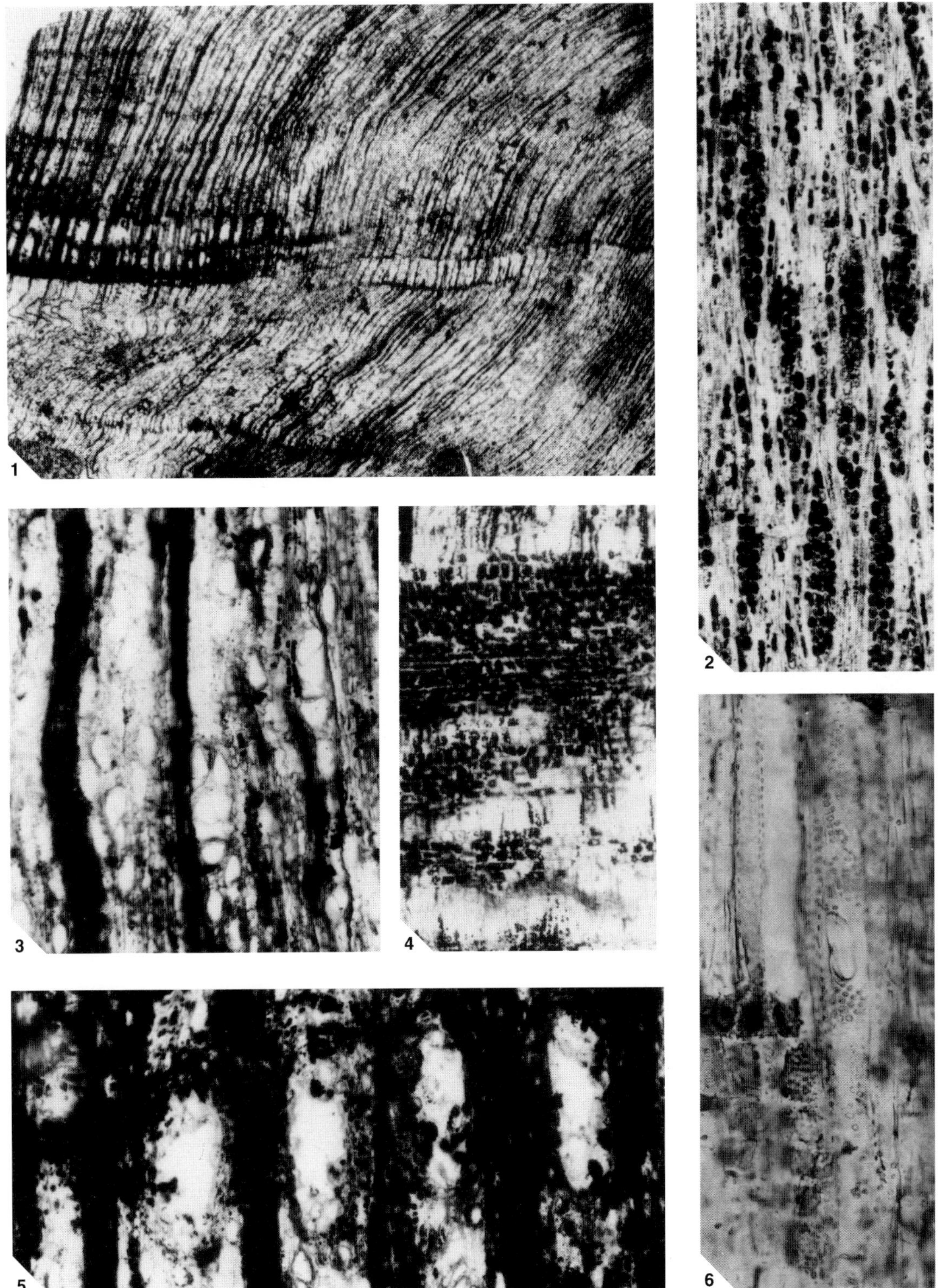

CRAWLEY, *Euphorbioxylon*

Type species. Euphorbioxylon speciosum Felix, 1887, Tertiary?, Columbia.

Euphorbioxylon hernense sp. nov.

Plate 6

Derivation of name. After Herne Bay, Kent.

Holotype. NHM V.794D; slides V.794D$1–8.

Locality and horizon. Reculvers, Herne Bay, Kent; Palaeocene Thanet Sand Formation.

Diagnosis. Growth rings, indistinct to absent. Vessels diffuse porous, mean density of 58 mm^2, solitary (67%) and in radial multiples of 2–3, mean tangential diameter 30 μm, perforation plates simple, vessel to vessel pitting alternate, bordered, 4–6 μm, vessel to parenchyma pitting similar to vessel to vessel, minute, 2–3 μm. Axial parenchyma absent. Rays of two sizes, uniseriate and 4–5 seriate, mean height, multiseriate rays 430 μm or 27 cells, uniseriate rays 175 μm, heterogeneous, multiseriate ray margins of 1–2 square to upright cells. Libriform fibres are present. Traumatic axial canals in tangential series present or absent.

Description. Single piece of silicified secondary wood. Growth ring boundaries indistinct [**2**] marked by flattened fibres or axial canals.

Vessel elements. Vessels diffuse porous [**5**] (Pl. 6, fig. 1) with solitary vessels 67% and radial groups of 2–3 vessels (Pl. 6, fig. 3). Simple perforation plates [**13**] (Pl. 6, fig. 6). Intervessel pits alternate [**22**] (Pl. 6, fig. 6), bordered; pit size small [**25**], 4–6 μm; vessel-parenchyma pitting similar to intervessel in shape [**30**] but size minute, 2–3 μm; tangential diameter of vessel lumina range 16–44 μm, mean 30 μm [**40**]. Density 47–70 mm^2, mean 58 [**49**]. Vessel element length range 88–209 μm, mean 140 μm [**52**].

Imperforate tracheary elements. Libriform fibres, no pitting seen [**61**].

Axial parenchyma. None observed [**75**].

Ray parenchyma. Ray width 1–6 cells [**98**]; height of multiseriate rays range 275–770 μm (17–48 cells), mean 430 μm (27 cells); uniseriate rays range 88–297 μm (4–14 cells), mean 175 μm (eight cells). Rays of two distinct sizes [**103**] (Pl. 6, fig. 2), one and 4–6 cells wide. Width of multiseriate rays 33–66 μm (3–6 cells); width range of uniseriate rays 11–15 μm. Composition is heterogeneous (Pl. 6, fig. 4); body ray cells procumbent with 1–2 rows of upright marginal cells [**106, 107**]. Density 12–21 rays mm^2, mean 16 [**116**].

Intercellular canals. In tangential rows, probably of traumatic origin [**131**]; individual canals elliptical 88 × 330 μm (Pl. 6, figs 1, 5).

Comparisons. The apparent absence of axial parenchyma in this specimen together with rays of two sizes, small to minute alternate vessel pitting, occurs in some Recent woods of Celastraceae, Euphorbiaceae and Flacourticaeae. Large axial intercellular canals in tangential series are also present in the fossil (Pl. 6, fig. 1) and are only recorded in Euphorbiaceae. Within Euphorbiaceae, *Glochidion*-type wood lacks parenchyma but has separate fibres and multiseriate rays with margins taller than the body of the ray. In the Bridelieae (*Bridelia* and *Cleistanthus*) ray structure is similar to that of the fossil but the rays are not of two sizes and axial paratracheal parenchyma with septate fibres are present. Two euphorbiaceous fossil species are closely comparable to the Herne Bay wood: *Bridelioxylon fibrosum* Mädel, 1962 and *Securinegoxylon biseriatum* Mädel, 1962, both from the Upper Cretaceous of South Africa. *Securinegoxylon* is the most similar but differs in possessing septate fibres, rays that are only 1–3 seriate and larger vessel to parenchyma pitting. The Herne Bay wood is therefore distinct and included in the morphotaxon *Euphorbioxylon* for woods of euphorbiaceous affinity but not closely comparable to any Recent genus.

Remarks. This is the first euphorbiaceous fossil to be recorded from the Thanet Sand Formation. Fruits and seeds of Euphorbiaceae are first recorded from the Woolwich and Reading Formation and many species are known from the London Clay Formation (Chandler 1964). Large axial canals in tangential series are usually of pathological origin and caused by damage to the tree (Brazier and Franklin 1961). There is evidence of extensive termite attacks on other woods (*Edenoxylon*, *Cantia*) from this horizon.

Family FAGACEAE Dumortier, 1829

Morphotaxon CASTANOXYLON Navale, 1964

Type species. Castanoxylon indicum Navale, 1964, Upper Pliocene Cuddalore Series, Pondicherry, India.

Castanoxylon philpii sp. nov.

Plate 7

Derivation of name. After Mr E. G. Philp, former Curator of Maidstone Museum and Art Gallery, Kent.

Holotype. MM MNE/G/00004; slides MM MNE/G/00004$1–3.

Locality and horizon. Details unknown but probably Maidstone or Medway area, Kent; Palaeocene Thanet Sand Formation.

Diagnosis. Growth rings distinct. Vessels diffuse to semi-ring porous, radial to diagonal alignment, exclusively solitary, mean tangential diameter 160 μm. Mean vessel element length 410 μm. Perforation plates simple. Vessel to ray parenchyma pitting large, simple, elliptical in shape and almost vertically (axially) aligned. Mean ray density 26 mm^2, all rays exclusively uniseriate, mean ray height 265 μm or 16 cells, distinctly heterogeneous. Imperforate tracheary elements are vasicentric tracheids and fibres with bordered pits.

Description. Single silicified piece of secondary xylem. Growth ring boundaries distinct [**1**] marked by flattened latewood fibres and smaller vessel diameter of latewood vessels.

Vessel elements. Vessels semi-ring porous [**4**] to diffuse [**5**] (Pl. 7, fig. 1) with a radial to diagonal pattern [**7**]. Exclusively solitary [**9**] simple perforation plates [**13**]; vessel-parenchyma pitting simple and vertical (palisade) [**32**] (Pl. 7, fig. 3); pit size 15 μm; tangential diameter of vessel lumina range 78–222 μm, mean 160 μm [**42**]. Density 5–13 mm^2, mean 8 mm^2 [**47**]. Vessel element length range 253–550 μm, mean 410 μm [**53**].

Imperforate tracheary element. Vasicentric tracheids present [**60**] (Pl. 7, fig. 4); ground tissue fibres with distinctly bordered pits [**62**], thin to thick walled [**69**].

Axial parenchyma. Apotracheal, diffuse-in-aggregates [**77**] as short discontinuous, tangential bands.

Ray parenchyma. Exclusively uniseriate rays [**96**] (Pl. 7, fig. 5); height range 110–495 μm (6–30 cells), mean 265 μm (16 cells). Composition is heterogeneous (Pl. 7, fig. 2); rows of procumbent, square and upright cells present. Density 19–33 rays mm^2, mean 26 [**116**].

Comparisons. The following combination of features is found in this fossil: vessels solitary; presence of vascular tracheids; apotracheal diffuse-in-aggregates axial parenchyma; ray parenchyma that is exclusively uniseriate; and vessel to parenchyma pitting large, simple and vertically elongate (palisade). This is closely comparable to Recent *Castanopsis* and *Castanea* (Fagaceae). The Maidstone fossil is closest to *Castanopsis* according to data produced by Selmeier (1970) because it has radially arranged vessels.

The morphotaxon *Castanoxylon* incorporates *Castanopsis*-like wood but all species are distinctly ring-porous except *Castanoxylon eschweilerense* Burgh, 1973. *C. eschweilerense* differs from the Thanet wood by possessing paratracheal parenchyma, libriform fibres and homocellular to only weakly heterogeneous ray parenchyma; hence the erection of a new species, *C. philpii*, here.

Remarks. A *Castanea*-like wood was recorded by Tomkeieff and Blackburn (1942) from Rhum, Inner Hebrides, but was not described. It was possibly from the interbasaltic beds of the Palaeocene/Eocene boundary. *Castanopsis*? pollen grains have been recorded from the London Clay and Oligocene formations (Chandler 1964).

Family FAGACEAE Dumortier, 1829, PLATANACEAE? Dumortier, 1829, or ICACINACEAE? (Bentham) Miers, 1851

Morphotaxon HYTHIA Stopes, 1915

Type species. Hythia elgarii Stopes, 1915 from near Maidstone, Kent, England and Reculvers, Herne Bay, Kent, England; Palaeocene Thanet Sand Formation.

Hythia elgarii Stopes, 1915

Text-figure 5; Table 6

1915 *Hythia elgari* Stopes, p. 277, pls 29–30, text-figs 85–86.
1924 *Hythia elgari* Stopes; Scott, p. 54.
1931 *Hythia elgari* Stopes; Edwards, p. 46.
1971 *Hythia elgari* Stopes; Pant and Kidwai, p. 252.
1986 *Hythia elgari* Stopes; Süss, p. 176.

Holotype. NHM V.13232; slides NHM V.13232a-1 and MM MNE/G/00013.

Paratype. NHM V.794A; slides NHM V.794A$1–3.

Locality and horizon. Near Maidstone, Kent, England (V.13232, MM MNE/G/00013) and Reculvers, Herne Bay, Kent, England (V.794A); Palaeocene Thanet Sand Formation (see introduction to the Thanet Sand Formation for a discussion of the original provenance of *Cantia* and *Hythia*).

Diagnosis. Vessels diffuse porous, mean density 71–86 mm^2, mainly solitary and radial or tangential groups of 2–4, perforation plates exclusively scalariform with 8–15 bars, pitting transitional. Ray density 9–14 mm^2, dimorphous, uniseriate and 8–10 or more cells wide, multiseriate ray mean height 1641–2755 μm or 75–118 cells, width 127–232 μm or 10–12 cells, uniseriate rays common, heterogeneous with ray margins of 1–2 upright cells.

Description. A small piece of silicified secondary wood. Growth rings are absent [2].

Vessel elements. Vessels diffuse porous [5] (Text-fig. 5A–C); perforation plates not seen; vessel to vessel pitting transitional [**20–21**]; vessel parenchyma pitting similar [**30**]. Tangential diameter of vessel lumina range 22–44 μm, mean 34 μm [**40**]. Vessel density range 74–105 mm^2, mean 86 mm^2 [**49**].

Imperforate tracheary elements. Fibre-tracheids with distinctly bordered pits [**62**].

Axial parenchyma. Apotracheal, diffuse [**76**] and diffuse-in-aggregates [**77**] (Text-fig. 5C).

Ray parenchyma. Ray width 1–16 cells [**99**]; two distinct sizes [**103**], uniseriate and eight or more seriate. Multiseriate ray height range 1456–3780 μm (60–162 cells), mean 2755 μm (118 cells) [**102**]; uniseriate rays could not be measured accurately. Composition is heterogeneous; ray margins composed of 1–2 square cells; uniseriate rays composed of upright cells 9–18 mm^2, mean 14 mm^2.

Comparisons. Stopes (1915) made a tentative but reasonable comparison of *Hythia* to the wood of *Fagus*

EXPLANATION OF PLATE 7

1–5. *Castanoxylon philpii* sp. nov., probably Maidstone or Medway area, Kent, Palaeocene Thanet Sand Formation, holotype, MM MNE/G/00004. 1, TS, apotracheal diffuse-in-aggregates parenchyma can be seen across the centre, MM MNE/G/00004$1; ×24. 2, RLS, heterogeneous rays, MM MNE/G/00004$3; ×240. 3, RLS, vessel to ray pitting, MM MNE/G/00004$3; ×500. 4, TS, vasicentric tracheids can be seen as a zone of elements surrounding some vessels, MM MNE/G/00004$1; ×200. 5, TLS, exclusively uniseriate rays, MM MNE/G/00004$2; ×200.

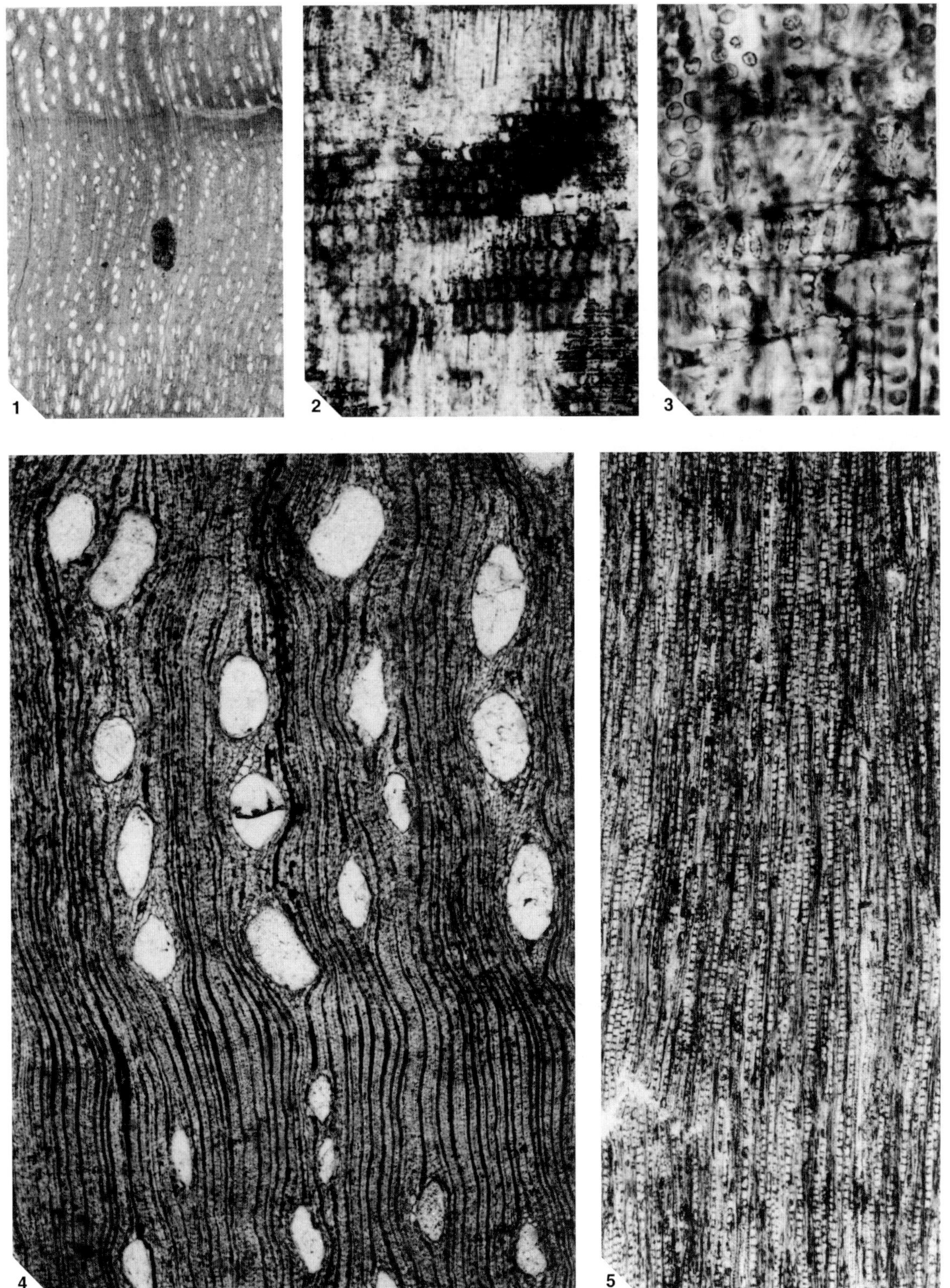

CRAWLEY, *Castanoxylon*

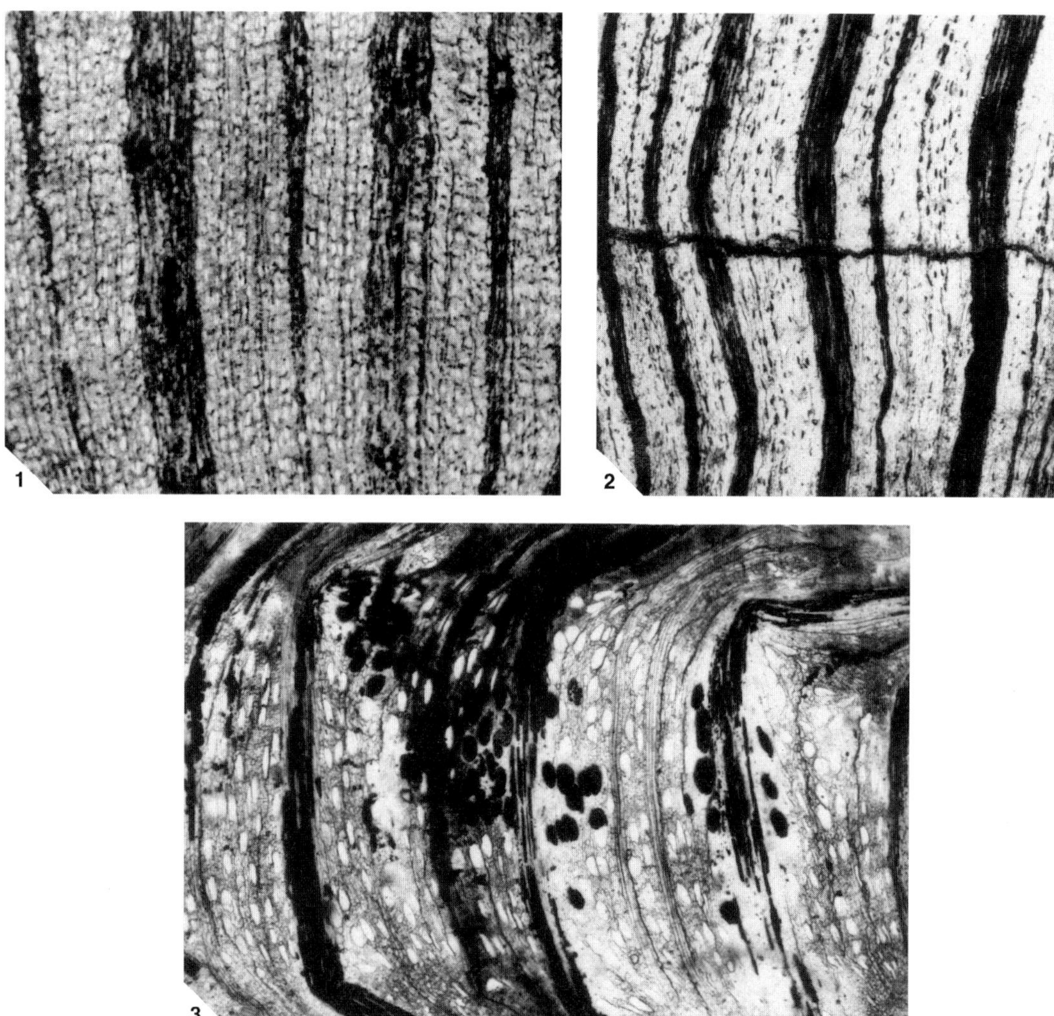

TEXT-FIG. 5. *Hythia elgarii* Stopes, 1915. A, Herne Bay, Kent, England, Palaeocene Thanet Sand Formation. B–C, near Maidstone, Kent, England, Palaeocene Thanet Sand Formation. A, TS, showing very wide multiseriate rays with uniseriate rays in between, paratype, NHM V.794$1; ×100. B, TS, as 1, holotype, NHM V.13232c; ×100. C, TS, showing characteristic dark plugging (cf. Pl. 5, figs 1–2), holotype, NHM V.13232b; ×246.

(Fagaceae). *Hythia*-type wood with uniseriate rays composed of upright cells, multiseriate rays at least ten cells broad and often over 3 mm high, exclusively scalariform perforation plates, and transitional pitting also occurs in the Eupteleaceae, Icacinaceae and Platanaceae. This problem was recognised by Page (1968) who included fossil woods having solitary vessels, apotracheal diffuse-in-aggregate parenchyma, wide multiseriate rays and scalariform perforation plates in *Plataninium*. In a comprehensive review of *Fagus*-like fossil wood, Süss (1986) excluded *Hythia* from Fagaceae. He thought that the structure was far closer to *Icacinoxylon* Schilkina, 1956, a genus erected to include wood structure like that of Recent unspecialized Icacinaceae, e.g. *Citronella* and *Ottoschultzia*. Wheeler *et al.* (1995) noted that there is a need to re-evaluate all woods assigned to *Icacinoxlylon* and *Plataninium* because the boundaries of the two genera are not clearly defined. To this I would add *Hythia*, which should be kept as a separate morphotaxon until a revision is done.

Both *Icacinoxylon laticiphorum* Greguss, 1969 and *I. cristallophorum* Greguss, 1969 from the Oligocene of Hungary show similar structure to *Hythia elgarii*. However, they differ in two important features: *Icacinoxylon laticiphorum* has radial canals in the multiseriate rays whereas *I. cristallophorum* has highly crystalliferous rays.

Remarks (see also remarks on *Cantia arborescens*). Although Stopes's description is good for most qualitative features there is a lack of quantitative data. Also, she did not mention that uniseriate rays, a diagnostic character, are present between the broad multiseriate rays (Pl. 4, figs 1–2). The uniseriate rays appear to be composed entirely of upright cells but their height is impossible to measure in both specimens. As can be seen from Table 6, the specimens are similar for all features and most numerical data. The variation in vessel diameter, ray density, width and height are within acceptable intraspecific limits for this type of wood (see Crawley 1989 for remarks on variation within specimens of the structurally similar *Plataninium decipiens* Brett, 1972).

Icacinaceous remains have been recorded from the Thanet Sand Formation (Collinson 1983), and Woolwich and Oldhaven Beds, with abundant seed taxa in the London Clay Formation (Chandler 1964).

Discussion of woods from the Thanet Sand Formation

All of the specimens of silicified wood from Herne Bay were collected loose on the beach and not *in situ*. In the past they have been attributed to the Thanet Formation ('Corbula' and Morrisi Beds) or the Woolwich and Reading Formation (Brett 1966), usually because they were found over that particular bed on the beach. Ward (1978) listed silicified wood as occurring in the following: Thanet Formation, Unit C (*Arctica morrisi* Bed); Unit E (*Astarte tenera* Bed); Unit F (Tornatellaea Parisians Bed); Woolwich and Reading Formation Unit J (Beltinge Fish Bed) and possibly Woolwich Marine Bed Unit K. Brett (1966) analyzed sediment infilling a Teredo boring within *Edenoxylon*. He concluded that it matched sand in the lower Woolwich Beds. However, study of fillings in *Cantia, Ilicoxylon* and the new specimen of *Edenoxylon* revealed a Thanet Sand Formation-type sediment of pale, consolidated sand with some glauconite grains (J. Cooper and D. W. Ward, pers. comm. 1990). A re-examination of Bretts' *Edenoxylon* showed no usable sediment but his specimen of *Quercinium porosum* Brett, 1960 did. This sediment was again identified as typical of the Thanet Sand Formation (J. Cooper and D. W. Ward, pers. comm. 1990).

Edenoxylon aemulum was described by Brett as essentially tropical in nature. The presence of semi-ring porosity in this wood does not invalidate this but suggests some seasonality of climate (Gilbert 1940). This type of growth-ring is classified as Type 5 and associated with gradual seasonal change (Carlquist 1988). There is some evidence of incremental growth in *Cantia arborescens* but it is vague or appears irregular, and perhaps represents 'false' growth rings (non-seasonal periodicity). Extant Betulaceae are mainly temperate but do have a few subtropical representatives. *Ilicoxylon prestwichii* is similar in qualitative features to extant tropical montane species of *Ilex* (Baas 1973). These woods are very similar to their temperate relatives but lack spiral or helical sculpture in the vessels or fibres (as does *I. prestwichii*). According to Brett (1960), *Quercinium porosum* has structure similar to that found today in species of evergreen oaks from the tropics, i.e. absent or indistinct growth-rings and diffuse porous vessels. Further evidence comes from two coniferalean woods from the Thanet Sand Formation. One (V.4957) has growth rings but the latewood is not conspicuous, whereas the other, *Ginkgoxylon* sp. (Text-fig. 6A–B; identification by Dr D. Pons, pers. comm. 1989), has very vague evidence of growth rings, which are probably not seasonal (G. T. Creber, pers. comm. 1989). Calculation of vulnerability indices (Carlquist 1977) for all the Herne Bay dicotyledonous woods are included in Table 15. They are compared with the other British Palaeocene–Lower Eocene woods. Seasonality of climate is perhaps indicated by the very low figure for *Cantia*. However both *Cantia* and *Ilicoxylon* have long, scalariform perforation plates, a feature associated with trees of moist, relatively non-seasonal forest in tropical to temperature regions (Carlquist 1988). In conclusion, the woods seem to indicate subtropical to tropical conditions with some seasonality.

TABLE 6. *Hythia elgarii*: comparison of the holotype with another specimen from the Thanet Sand Formation.

		V.794A	V.13232
Growth rings		not present	not present
Vessels			
Distribution		diffuse porous	diffuse porous
Grouping		mostly solitary and in tangential and oblique groups of 2–4	mostly solitary and in tangential and oblique groups of 2–3
Density/mm^2			
	range	74–105	60–82
	mean	86	71
Tangential diameter μm			
	range	22–44	22–77
	mean	34	45
Element length μm			
	range	–	–
	mean	–	–
Perforation plate		–	exclusively? scalariform with 8–15 bars
Pitting			
	intervascular	transitional	transitional
	vessel to parenchyma	similar	similar
Parenchyma			
		apotracheal diffuse and diffuse-in-aggregates	apotracheal diffuse and diffuse-in-aggregates
Rays			
Density/mm			
	range	9–18	7–13
	mean	14	10
		rays of two sizes, 1 and 8 or more cells wide; large rays often dissected	rays of two sizes, 1 and 10 or more cells wide
Multiseriate ray width μm			
	range	110–330	21–224
	mean	232	127
Multiseriate ray width cells			
	range	8–16	2–19
	mean	12	10
Multiseriate height μm			
	range	1456–3780	504–3780
	mean	2755	1641
Multiseriate height cells			
	range	60–162	23–171
	mean	118	75
Uniseriate height μm			
	range	–	–
	mean	–	–
Composition		heterocellular, multiseriate; ray margins of 1–2 square cells; uniseriate rays common and composed of upright cells; multiseriate ray cells up to 60 μm high; uniseriate ray cells 55–120 μm high	probably heterocellular type 2B, multiseriate ray margins 1–2 possibly square or upright cells; uniseriate rays common, composed of upright cells; multiseriate ray cells 22–35 μm high, uniseriate ray cells 60–75 μm high
Imperforate tracheary elements		fibre-tracheids	fibre-tracheids

Woods from the Reading Formation

The Reading Formation is perhaps best known for its rich leaf and infructescence remains (Collinson and Crane 1978; Crane 1978, 1981, 1984; Crane and Manchester 1982; Herendeen and Crane 1992). These occur as compressions in lacustrine or river deposits within clayey lenses interbedded with sands. Silicified wood also occurs but has been little studied (Crawley 1989). This is probably because the compression flora is more immediately accessible to researchers and currently, the horizon (or horizons) from which the wood comes has yet to be identified. Most of the wood described herein comes from the Hermitage Pit near Newbury, Berkshire and was collected loose. Recent collecting at Hermitage has still not resolved this matter (Mr L. R. Lewis and Ms J. Skipper, pers. comm. 1999). Data on wood supplements those derived from the compression flora. Crawley (1989) described platanaceous, sterculiaceous and a possible lauraceous wood from this formation. The lauraceous wood was possibly found *in situ*.

Family ANACARDIACEAE Lindley, 1830 or BURSERACEAE Kunth, 1824

Morphotaxon CANARIOXYLON Prakash, Brezinova and Bužek, 1974

Type species. Canarioxylon ceskobudejovicenese Prakash, Brezinova and Bužek, 1974, Oligocene?, south Bohemia, Czechoslovakia.

Canarioxylon chieveleyense sp. nov.

Plate 8; Table 7

Derivation of name. After Chieveley, Berkshire.

Holotype. NHM V.63224, V.63224$1–3; collected by Mr L. R. Lewis, 1988–1989.

Paratypes. NHM V.63225, V.63227–30; slides V.63225, V.63225$1–3, V.63227$1–3–30$1–3; collected by Mr L. R. Lewis, 1988–1989.

Other material. NHM V.63226, V.63231–33; slides V.63231$1–3–V.63234$1–3; collected by Mr L. R. Lewis, 1988–1989, except V.63232, collected by M. Crawley in 1990.

Locality and horizon. Hermitage Farm Pit, Chieveley, near Newbury, Berkshire (V.63224–33); Palaeocene Reading Formation.

Diagnosis. Growth rings absent. Vessels diffuse porous, mean density 15–22 mm^2, mean tangential diameter range 101–121 μm, mean element length 266–481 μm, perforation plates simple, vessel to vessel pitting, alternate, bordered, large, 8–11 μm, vessel to parenchyma pitting simple to reduced borders, round to elliptical, tangential diameter 5·5–20 μm, tyloses common. Axial parenchyma paratracheal, vasicentric, sheaths one cell thick. Ray mean density range 9–17 mm^2, width 1–4 cells but usually 2–3 cells wide, mean ray height range, multiseriate 295–366 μm or 15–18 cells, uniseriate 127–213 μm or 4–6 cells, heterogeneous with a single marginal row of upright cells. Septate fibres are present.

Description. (See Table 7 for full quantitative data). Six pieces of silicified secondary wood. Growth ring boundaries absent [**2**].

Vessel elements. Vessels diffuse porous [**5**] (Pl. 8, fig. 1); solitary 40–55% and with radial vessel groups of mostly 2–6. Vessels have simple perforation plates [**13**]; intervessel pits alternate, bordered; pit size medium [**26**] to large [**27**], 8–11 μm (Pl. 8, fig. 6); vessel to parenchyma pits with reduced borders to simple, rounded to oval [**31**] (Pl. 8, fig. 4). Tangential diameter of vessel lumina 101–121 μm [**42**]. Density 15–22 mm^2 [**47–48**]. Vessel element length 266–481 μm [**52–53**]; tyloses common [**56**].

Imperforate tracheary elements. Septate fibres [**65**] (Pl. 8, fig. 2), very thin walled [**68**].

Axial parenchyma. Paratracheal, vasicentric [**79**] (Pl. 8, fig. 3), sheath one cell thick.

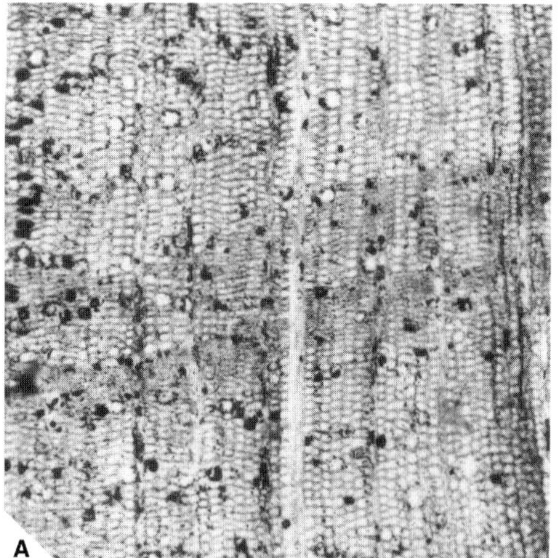

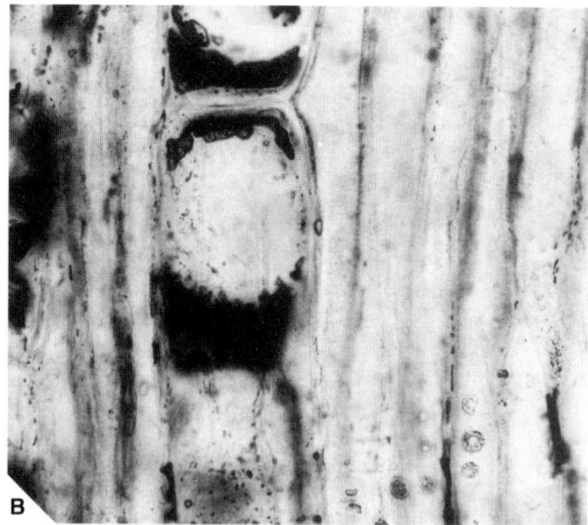

TEXT-FIG. 6. *Ginkgoxylon* sp., Herne Bay, Kent, England, Palaeocene Thanet Sand Formation, NHM V.17079. A, TS, vague evidence of a growth ring, NHM V.17079$1; ×50. B, RLS, large axial parenchyma cells characteristic of *Ginkgo*-type wood; the cells are huge in comparison with the fibre-tracheids to right of figure, NHM V.17079$2; ×500.

Ray parenchyma. Ray width 1–4 cells [**97**]; multiseriate ray height 295–366 μm (15–18 cells); uniseriate ray height 127–213 μm (4–6 cells); ray width 28–38 μm (three cells). Composition is heterogeneous (Pl. 8, fig. 2); body ray cells procumbent with one row of rectangular to upright marginal cells (Pl. 8, fig. 5) [**106**]. Density 9–17 rays mm^2 [**115–116**].

Comparisons. The following combination of features are found in these woods: vessels solitary and in radial groups of usually 2–3; perforation plates exclusively simple; vessel to vessel pitting alternate, bordered and medium to large, vessel to parenchyma pitting simple to part bordered, round to oval; axial parenchyma paratracheal, vasicentric; rays usually 2–3 seriate, heterogeneous with a single marginal row of upright cells; and septate fibres. This combination is found today in Anacardiaceae, Burseraceae, Euphorbiaceae and Lauraceae. Of these the Anacardiaceae and Burseraceae are probably most similar. Euphorbiaceae usually have markedly heterogeneous rays with many marginal rows (except in the Brideliae) and longer radial groups. Lauraceae often have idioblasts in axial or ray parenchyma or both. Fossil woods with the most similar structure are: *Burseroxylon garugoides* Lakhanpal, Prakash and Awasthi, 1981, from the Tertiary of India; *B. preserratum* Prakash and Tripathi, 1975, from the Palaeocene/Eocene of India; *B. preserratum* Prakash and Tripathi, 1975, *in* Bande and Prakash 1983, from the Tertiary of India; and *Canarioxylon indicum* Ghosh and Roy, 1978, also from the Tertiary of India. The Newbury woods can be distinguished from *Burseroxylon garugoides* Lakhanpal, Prakash and Awasthi, 1981 and *B. preserratum* Prakash and Tripathi, 1975, because they possess crystalliferous rays, whilst the most similar, *Canarioxylon indicum* Ghosh and Roy, 1978, can be separated by its much larger tangential vessel diameters and longer vessel elements. The Reading Formation woods are regarded as a new species and placed in *Canarioxylon*.

EXPLANATION OF PLATE 8

1–6. *Canarioxylon lewisii* sp. nov., Hermitage Farm Pit, Chieveley, near Newbury, Berkshire, Palaeocene Reading Formation. 1, TS, holotype, NHM V.63224$1; ×24. 2, TLS, note septate fibres between rays, paratype, NHM V.63229$2; ×120. 3, TS, note paratracheal vasicentric axial parenchyma, paratype, NHM V.63229$1; ×60. 4, RLS, vessel to axial parenchyma pitting, paratype, NHM V.63229$2; ×500. 5, RLS, heterogeneous rays, holotype, NHM V.63224$3; ×120. 6, TLS, vessel to vessel pitting, paratype, NHM V.63229$2; ×500.

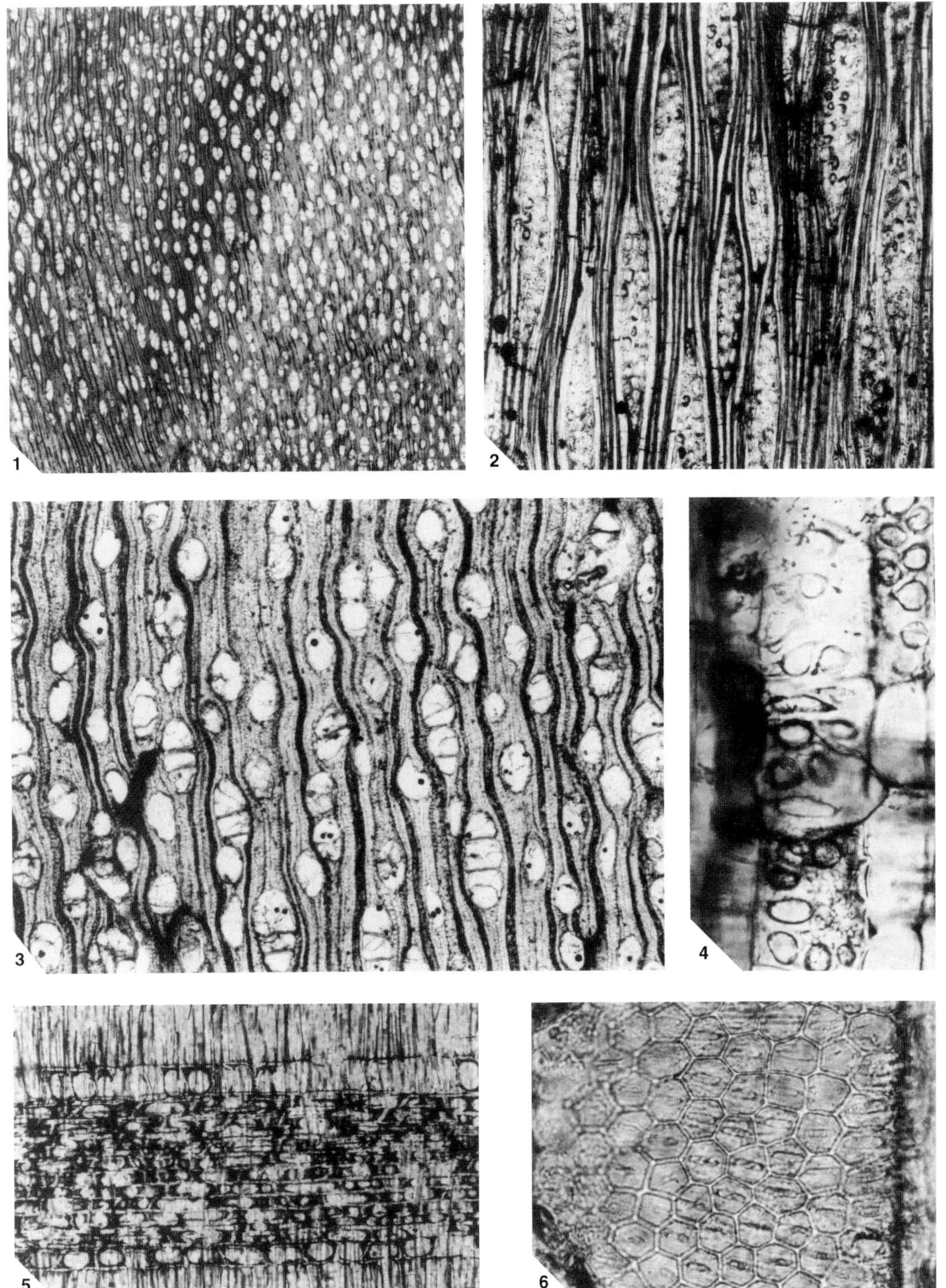

CRAWLEY, *Canarioxylon*

TABLE 7. *Canarioxylon chieveleyense* sp. nov.: comparison of quantitative data.

	V.63224	V.63225	V.63227	V.63228	V.63229	V.63230
Vessels						
tangential diameter						
mean (μm)	107	112	109	108	12	101
range	33–154	66–154	44–165	55–154	6–165	44–143
element length						
mean (μm)	393	–	–	481	323	266
range	187–517	–	–	33–605	220–462	121–429
density/mm^2						
mean	21	16	21	22	19	15
range	11–34	10–23	13–30	16–29	11–31	9–23
solitary vessels %						
mean	54	40	55	55	56	55
range	46–61	39–40	54–55	54–56	52–58	51–58
radial groups	2–4	2–3	2–4	2–6	2–4	2–3
Ray parenchyma						
multiseriate height						
mean (μm)	331	–	–	361	366	295
range	132–462	–	–	209–550	165–803	176–451
mean (cells)	17	–	–	18	18	15
range	7–23	–	–	10–28	8–40	9–23
multiseriate width						
mean (μm)	32	–	–	32	38	28
range	11–44	–	–	11–44	14–55	11–39
mean (cells)	2	–	–	2	3	2
range	1–3	–	–	1–3	1–4	2–6
uniseriate height						
mean (μm)	146	–	–	213	152	127
range	88–242	–	–	88–407	88–275	66–209
mean (cells)	2	–	–	6	3	4
range	1–3	–	–	3–12	1–4	2–6
ray density per tangential mm						
mean	17	10	12	9	11	12
range	12–22	9–12	9–17	6–12	8–16	8–16

Remarks. No other anacardiaceous or burseraceous remains have been recorded from the Reading Formation but fruits and pollen are known from the London Clay Formation (Chandler 1964).

Family ANACARDIACEAE Lindley, 1830, BURSERACEAE Kunth, 1824, EUPHORBIACEAE A. Jussieu, 1789 or LAURACEAE Jussieu, 1789

Morphotaxon PARAPHYLLANTHOXYLON Bailey, 1924

EXPLANATION OF PLATE 9

1–5. *Paraphyllanthoxylon chievleyense* sp. nov., Hermitage Farm Pit, Chieveley, both near Newbury, Berkshire, Palaeocene Reading Formation. 1, TS, paratype NHM V.63242$1; ×24. 2, TLS, vessel to vessel pitting, holotype NHM V.63235$2; ×480. 3, RLS, vessel to ray pitting, holotype, NHM V.63235$3; ×480. 4, TS, notice lack of axial parenchyma, holotype, NHM V.63235$1; ×60. 5, TLS, exclusively multiseriate rays, holotype, NHM V.63235$2; ×60.

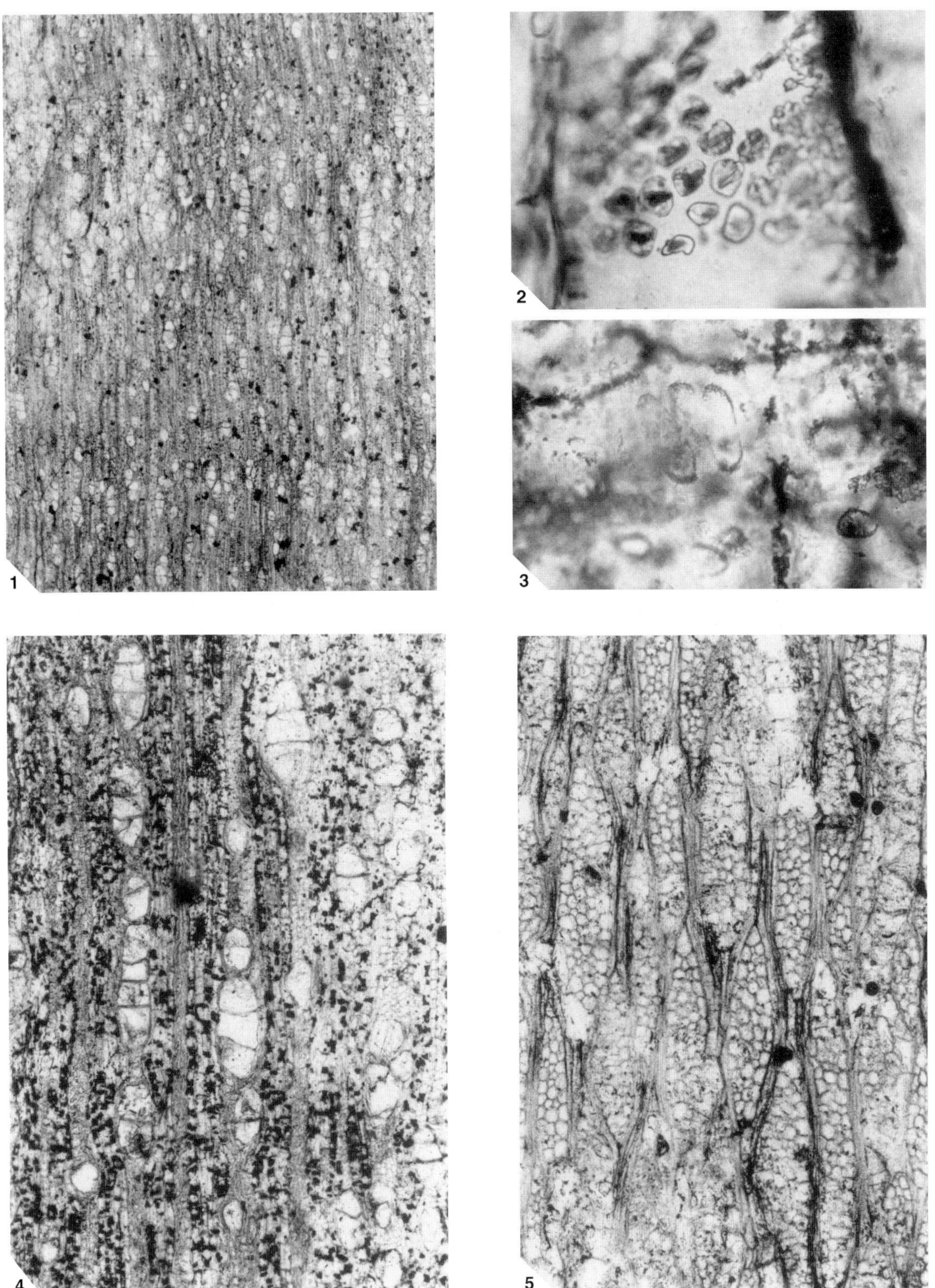

CRAWLEY, *Paraphyllanthoxylon*

TABLE 8. *Paraphyllanthoxylon lewisii*: comparison of quantitative data.

	V.24	V.63235	V.63236	V.63240	V.63241	V.63242
Vessels						
tangential diameter						
mean (μm)	168	141	131	146	125	138
range	56–212	50–198	66–187	55–209	55–187	99–187
element length						
mean (μm)	335	–	–	–	–	238
range	140–420	–	–	–	–	99–385
density/mm^2						
mean	9	11	10	12	10	10
range	7–11	9–15	7–13	9–16	8–13	7–12
solitary vessels %						
mean	60	36	34	28	37	37
range		29–35	28–40	27–29	36–40	22–51
radial groups	2–5	2–4	2–3	2–4	2–5	2–4
Ray parenchyma						
multiseriate height						
mean (μm)	900	831	838	808	767	744
range	400–1400	336–1456	392–1204	280–1288	364–1330	224–1344
mean (cells)	16	29	17	16	15	16
range	10–31	7–17	9–24	6–26	7–27	5–27
multiseriate width						
mean (μm)	84	138	125	133	148	134
range	50–120	56–196	56–82	63–196	77–196	56–196
mean (cells)	2	4	4	4	4	4
range	2–3	2–6	2–5	2–6	2–6	2–6
ray density per tangential mm						
mean (μm)	–	5	7	6	5	5
range	4–6	4–6	5–9	4–9	4–7	4–6

Type species. Paraphyllanthoxylon arizonense Bailey, 1924, Upper Cretaceous, Arizona, USA

Paraphyllanthoxylon lewisii sp. nov.

Plate 9; Table 8

1989 *Ulminium*? sp.; Crawley, p. 611, pl. 69, figs 7–8.

Derivation of name. After the collector of the holotype specimen, Mr L. R. Lewis.

Holotype. NHM V.63235; slides V.63235$1–3.

Paratypes. NHM V.24, V.63236, V.63240–42; slides V.24$1–6, V.63236$1–3, V.63240$1–3–42$1–3.

Other material. NHM V.63237–39, V.63243–44; slides V.63237$1–3–V.63239$1–3, V.63243$1–3–V.63244$1–3.

Locality and horizon. Shaw (V.24) and Hermitage Farm Pit, Chieveley, both near Newbury, Berkshire; Palaeocene Reading Formation.

Collection details. Collected by Professor T. R. Jones (V.24), Mr L. R. Lewis, 1988–1990 (V.63235, V.63244), and Mr A. Higgott, 1988–1989 (V.63236–43).

Diagnosis. Growth rings absent. Vessels diffuse porous, mean density 9–12 mm^2, mean tangential diameter 125–168 μm, mean element length range 238–335 μm, perforation plates simple, vessel to vessel pitting alternate, bordered, large, 9–14 μm, vessel to parenchyma simple to reduced borders, round and horizontally to vertically elongate, large, tyloses common. Axial parenchyma absent. Mean density of rays 5–7 mm^2, 2–6 cells but usually 3–5 cells wide, mean height 744–900 μm or 15–17 cells, heterogeneous, with a single marginal row of upright cells. Septate fibres are present.

Description. (See Table 8 for full quantitative data). Six pieces of silicified secondary wood. Growth ring boundaries absent [**2**].

Vessel elements. Vessels diffuse porous [**5**] (Pl. 9, fig. 1); solitary 28–60% and in radial vessel groups of 2–5. Simple perforation plates [**13**]; intervessel pits alternate, bordered; pit size medium [**26**] to large [**27**] (Pl. 9, fig. 2), 9–14 μm. Vessel to parenchyma pits with reduced borders to simple, horizontal (gash-like) to vertical (palisade) [**32**] (Pl. 9, fig. 3); tangential diameter of vessel lumina 125–168 μm [**42**]. Density 9–12 mm^2 [**47**]. Vessel element length 238–335 μm [**52–53**]; tyloses common [**56**].

Imperforate tracheary elements. Septate fibres [**65**] very thin walled [**68**].

Axial parenchyma. Absent [**75**] (Pl. 9, fig. 4).

Ray parenchyma. Ray width 2–6 cells [**97–98**] (Pl. 9, fig. 5); multiseriate ray height 744–900 μm (15–17 cells), uniseriate rays absent; ray width 84–148 μm (3–5 cells). Composition is heterogeneous; body ray cells procumbent with one row of square to upright marginal cells [**106**]. Density 5–7 rays mm^2 [**115**].

Comparisons. The following combination of features is found in these woods: vessels solitary and in radial groups of usually 2–3; perforation plates exclusively simple; vessel to vessel pitting alternate, bordered and medium to large; vessel to parenchyma pitting simple to part bordered, round to horizontally or vertically elongate; axial parenchyma absent; rays usually 2–6 seriate, uniseriate rays absent, heterogeneous with a single marginal row of upright cells; septate fibres. This combination is found in Recent Anacardiaceae, Buseraceae, Euphorbiaceae and Lauraceae. Of these, Burseraceae is most similar with more genera lacking axial parenchyma (Wheeler 1991). Euphorbiaceae usually have markedly heterogeneous rays with many marginal rows (except in the Brideliae), more radial groups with more numerous vessels, some axial parenchyma and common uniseriate rays. Lauraceae often have idioblasts in axial or ray parenchyma or both, and some paratracheal parenchyma. Anacardiaceae usually have some paratracheal scanty parenchyma (Wheeler 1991). Described woods with a similar structure are: *Canarioxylon* sp. from the Oligocene? of Czechoslovakia (Prakash *et al*., 1974); *Paraphyllanthoxylon anasazi* Wheeler, McClammer and LaPasha, 1995 and *P. arizonense* Bailey, 1924 from the Upper Cretaceous and Palaeocene of New Mexico, USA (Wheeler *et al*., 1995); *Paraphyllanthoxylon abbottae* Wheeler, 1991 from the Palaeocene of the USA; *P. capense* Mädel, 1962 from the Upper Cretaceous of the Republic of South Africa; *P. idahoense* Spackman, 1948 from the Lower Cretaceous of the USA; *P. utahense* Thayn, Tidwell and Stokes, 1983 from the Lower Cretaceous of the USA. Of these, *P. abbotae* is the nearest to the Newbury woods and is closely comparable in all qualitative features. It differs from *P. lewisii* as follows: it has larger vessels with mean ranges of between 162–234 μm compared to 125–168 μm; longer vessel-element length with mean ranges of 440–667 μm compared to 238–335 μm; and shorter multiseriate rays with mean ranges of 470–751 μm compared to 744–900 μm. The consistent nature of these differences is further supported by the sample size (*P. abbottae*, 30 specimens; *P. lewisii*, seven specimens), all of which represent mature secondary wood. Wheeler *et al*. (1995) stated that differences between *Paraphyllanthoxylon* species are primarily based upon quantitative features.

Remarks. Wheeler (1991) reviewed the morphotaxon *Paraphyllanthoxylon* and remarked on the homogeneity within the genus from Lower Cretaceous–Palaeocene deposits. It is also the most common type of dicotyledonous tree wood known from the Cretaceous and Palaeocene (Wheeler *et al*. 1995). *P. abbottae* was named after Dr Maxine Abbott. It was published as *P. abbottii* Wheeler, 1991, but the correct ending to the specific epithet should be feminine, i.e. '*ae*' as used herein.

Prakash *et al*. (1986) transferred the Cretaceous *Paraphyllanthoxylon alabamense* Cahoon, 1972, *P. idahoense* Spackman, 1948, and *P. utahense* Thayn, Tidwell and Stokes, 1983 to *Phyllanthinium*, and stated that this genus was related to Recent Phyllanthoideae (Euphorbiaceae). Wheeler (1991) was not convinced of such a relationship. She stated that the Paraphyllanthoxylon woods lack the high uniseriate ray margins and numerous uniseriate rays typical of Phyllanthoideae. She regarded Paraphyllanthoxylon as having a structural pattern represented today by the Anacardiaceae, Burseraceae, Euphorbiaceae, Flacourtiaceae and Lauraceae, but without positive relationships to a single family. This approach is followed for the Reading Formation specimens that are included here in Paraphyllanthoxylon.

Crawley (1989) described a specimen from the Reading Formation near Newbury as *Ulminium*? sp. because he thought that there were idioblasts at the ray margins, albeit poorly preserved. The new material is better preserved and there are no idioblasts present.

Family FAGACEAE Dumortier, 1829

Morphotaxon FAGOXYLON Stopes and Fuji, 1910, emend. Süss, 1986

Fagoxylon sp.

Text-figure 7

Material. NHM V.63172; slides V.63172$1–3.

Locality and horizon. Found loose on surface by Mr A. K. Knox near Woldingham, Surrey; Palaeocene Reading Formation.

Description. A large portion of silicified secondary wood with no primary wood. No distinct growth rings are present [**2**] but some evidence of possible incremental growth can be seen.

Vessels. Vessels diffuse porous [**5**] (Text-fig. 7A), mostly solitary [**9**] but some radial and tangential groups of up to four vessels occur; a single scalariform perforation plate [**14**] with at least nine bars [**15–16**] (Text-fig. 7B); pitting details are not well preserved but may be alternate in part [**22**]. Vessel element length indeterminate; tangential diameter of vessel lumina range 16–55 μm, mean 42 μm [**40**]. Vessel density range 80–126 mm^2, mean 104 mm^2 [**50**].

Imperforate tracheary elements. These appear to be fibre-tracheids as some bordered pits were seen [**61**] but they are generally very poorly preserved.

Axial parenchyma. Possible parenchyma strands were observed in LS occurring among the ground tissue, possibly apotracheal diffuse type [**76**].

Ray parenchyma. Dimorphous, the small rays are up to five cells wide and the large rays 11 or more cells wide [**99, 103**]; large rays often vertically aligned (Text-fig. 7C); large multiseriate ray height range 840–2660 μm [**102**] (30–110 cells), mean 1638 μm (60 cells); small multiseriate ray height range 140–644 μm (6–29 cells), mean 291 μm (13 cells); large ray width range 280–448 μm (11–21 cells), mean 363 μm (17 cells); small multiseriate ray width range 11–44 μm (1–5 cells), mean width 31 μm (three cells). Composition is homogeneous to subheterogeneous with multiseriate ray margins of 1–3 cells [**107**]; the uniseriate rays are composed of mainly procumbent cells. Ray density range 4–11 mm^2, mean 8 mm^2 [**115**].

Comparisons. The features of this wood are similar to those of extant *Fagus* spp. (beech, Fagaceae) or *Platanus* spp. (plane, Platanaceae). Typical features of *Fagus* secondary wood are: diffuse porous, small vessels with scalariform perforation plates; dimorphous rays with the smaller rays multiseriate and the larger rays often very high and broad; and fibre-tracheids present. *Platanus* secondary wood does not possess dimorphous rays or fibre-tracheids. The fossil possesses all the main features of *Fagus* and is, therefore, recorded as *Fagoxylon* Stopes and Fujii, 1910, emend. Süss, 1986, a morphotaxon for *Fagus*-like woods. Süss (1986) comprehensively reviewed all the fossil woods said to be similar to Recent *Fagus*. Unfortunately, because of the generally poor state of preservation, close comparison of the present fossil to these other species is impossible; hence, it is determined merely as *Fagoxylon* sp.

Remarks. The surface deposits in the Woldingham area consist in part of sub-Recent sands, the main

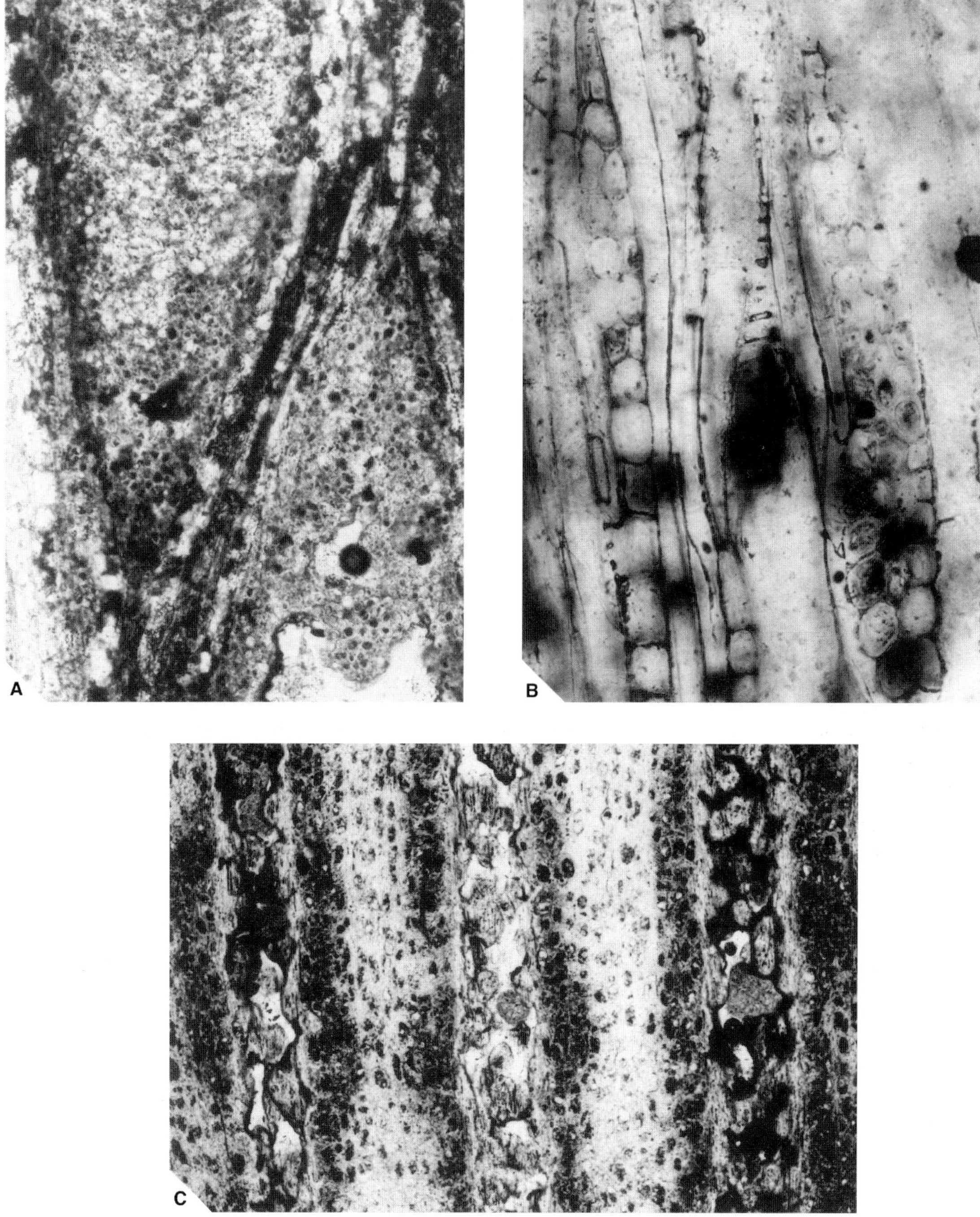

TEXT-FIG. 7. *Fagoxylon* sp., near Woldingham, Surrey; Palaeocene Reading Formation, NHM V.62172. A, TLS, large multiseriate rays with ends almost abutting, NHM V.63172$2; ×246. B, TLS, most well-preserved area showing in the centre a scalariform perforation plate flanked by almost homocellular rays, NHM V.63172$2; ×800. C, TS, diffuse porous vessels between large multiseriate rays, NHM V.63172$1; ×100.

constituent of which is thought to be Blackheath Beds (Dines and Edmunds 1933). A notable feature of the specimen is longitudinal fluting of the surface. *In situ* silicified wood collected mainly from the Reading Formation of the Newbury and Reading areas also has this characteristic fluting (which almost certainly occurred prior to fossilization). The colour and type of preservation are also consistent with such an origin. I have, therefore, little doubt in assigning this formation as the source of *Fagoxylon* sp.

Although some other woods are known from the Reading Formation (Crawley 1989), this is the first record of wood of the Fagaceae. Remains of probable Fagaceae, although not of *Fagus*, are known from the compression flora of the Reading Formation (P. Crane, pers. comm. 1990).

Today *Fagus* has a natural distribution that is restricted to temperate and subtropical regions of the Northern Hemisphere (Record and Hess 1943). However the Surrey *Fagoxylon*, unlike its Recent analogue, lacks growth rings and grew under aseasonal conditions.

Family MELIACEAE Jussieu, 1789

Morphotaxon ENTANDROPHRAGMINIUM Prakash, 1976

Type species. *Entandrophragminium aegyptiacum* Prakash, 1976, Tertiary (Eocene? Oligocene? Neogene?), Egypt.

Entandrophragminium lewisii sp. nov.

Plate 10

Holotype. NHM V.63245; slides V.63245$1–3.

Locality and horizon. Hermitage Farm Pit, Chieveley, near Newbury, Berkshire; Palaeocene Reading Formation; collected by Mr L. R. Lewis, 1988–1989.

Diagnosis. Growth rings absent. Vessels diffuse porous, mean density 6 mm^2, vessels solitary (65%) and in radial multiples of two vessels, mean tangential diameter 209 μm, perforation plates simple, vessel to vessel pitting alternate, bordered, 3–5 μm. Axial parenchyma paratracheal, vasicentric to confluent and locally banded, up to four cells broad. Ray mean density 5 mm^2, mean ray width 101 μm, range 2–7 cells, mean ray height 767 μm or 43 cells, homocellular to heterogeneous with one row of procumbent to square marginal cells. Solitary rhomboidal crystals in marginal ray cells.

Description. Single specimen of silicified secondary wood. Growth ring boundaries absent [**2**].

Vessel elements. Vessels diffuse porous [**5**] (Pl. 10, fig. 1); solitary 65% and in radial groups of two. Simple perforation plates [**13**]; intervessel pits alternate, bordered; pit size minute [**24**] to small [**25**], 3–5 μm; vessel to parenchyma pits not seen. Tangential diameter of vessel lumina range 99–275 μm, mean 209 μm [**43**]. Density 4–8 mm^2, mean 6 mm^2 [**47**]. Vessel element length range 341–396 μm; tyloses common [**56**] and thin walled.

Imperforate tracheary elements. Probably libriform fibres [**61**].

Axial parenchyma. Paratracheal, vasicentric [**79**], lozenge aliform [**81**], some confluent [**83**]; banded, up to four cells wide [**85**] (Pl. 10, fig. 3).

Ray parenchyma. Ray width 2–7 cells [**97**] (Pl. 10, fig. 2); multiseriate ray height range 363–1210 μm (21–67 cells), mean 767 μm (43 cells); uniseriate rays not seen; ray width range 44–132 μm (2–7 cells), mean 101 μm (five

---

EXPLANATION OF PLATE 10

1–5. *Entandrophragminium lewisii* sp. nov., Hermitage Farm Pit, Chieveley, both near Newbury, Berkshire, Palaeocene Reading Formation, holotype, NHM V.63245. 1, TS, NHM V.63245$1; ×24. 2, TLS, broad multiseriate rays with some sheath cells, NHM V.63245$2; ×100. 3, TS, showing paratracheal and banded parenchyma, NHM V.63245$1; ×60. 4, TS, possible axial canals in tangential series, NHM V.63245$1; ×60. 5, RLS, heterogeneous rays, NHM V.63245$3; ×180.

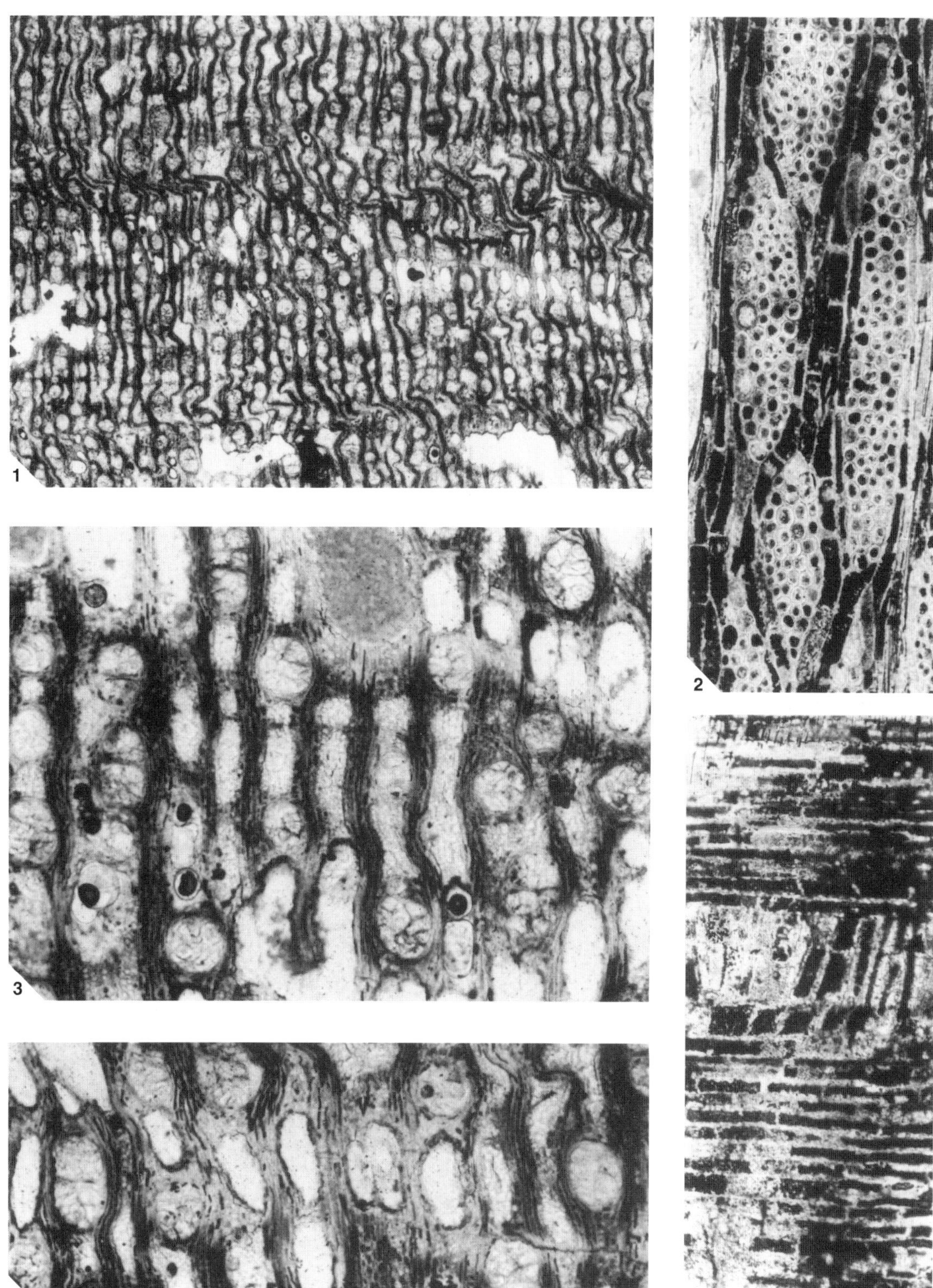

CRAWLEY, *Entandrophragminium*

cells). Composition is homocellular [**104**] to heterogeneous; body ray cells procumbent with one row of procumbent to square marginal cells [**106**]. Density 4–6 rays mm^2, mean 5 mm^2 [**115**]. Some sheath cells are present [**110**] (Pl. 10, fig. 2).

Intercellular canals. Possible axial canals of traumatic origin [**131**] (Pl. 10, fig. 4).

Mineral inclusions. Solitary prismatic crystals [**136**] in marginal ray cells [**137**] (Pl. 10, fig. 5).

Comparisons. The fossil has the following features: vessels large, solitary and in radial groups of two; small to minute pitting; axial parenchyma paratracheal vasicentric to confluent and banded; medullary rays commonly 2–6 cells in width and homocellular to weakly heterogeneous; solitary crystals in marginal ray cells. These features are found in Meliaceae and particularly the genus *Entandrophragma.* Fossil forms comparable to the Newbury wood are: *Carapoxylon ornatum* (Felix) Mädel, 1960 from the Upper Miocene of Germany; *C. cahenii* Lakhanpal and Prakash, 1970 from the Miocene of Africa; *Entandrophragminium magnieri* (Louvet) Prakash, 1976 from the Oligocene of Libya; *E. lateparenchimatosum* (Petrescu) comb. nov. from the Oligocene of Romania; and *E. normandii* (Louvet) comb. nov. from the Upper Eocene/ Lower Oligocene of Algeria. The closest species is *Entandrophragminium normandii* (Louvet) comb. nov., but this still differs from the Reading Formation wood in having radial groups of vessels of up to four, uniseriate rays and aliform parenchyma. The erection of a new species of *Entandrophragminium* (*E. lewisii*) is, therefore, justified.

Remarks. Two genera and species of fruit of Meliaceae have been recorded from the London Clay Formation (Chandler 1964).

Prakash (1976) revised the morphotaxon *Entandrophragmoxylon*, putting the type species, *E. boureauii* Louvet, 1963, into *Carapoxylon*, stating there was no substantial difference in the diagnoses. No mention was made of two other species described by Louvet (1968). Since then Petrescu (1978) has described a further species using the generic name *Entandrophragmoxylon*. Prakash (1976) instituted the morphotaxon *Entandrophragminium* for woods resembling Recent *Carapa* and *Entandrophragma*. All of the above occurrences of *Entandrophramoxylon* fall within Prakash's circumscription for *Entandrophragminium* rather than *Carapoxylon*, and are, therefore, transferred to it below.

Entandrophragminium lateparenchimatosum (Petrescu) comb. nov.

1978 *Entandrophragmoxylon lateparenchimatosum* Petrescu, p. 159, pls 59–60, text-figs 30–34.

Entandrophragminium mkrattaense (Louvet) comb. nov.

1968 *Entandrophragmoxylon mkrattaense* Louvet, pp. 103–104, pl. 13, text-figs 26–27.

Entandrophragminium normandii (Louvet) comb. nov.

1968 *Entandrophragmoxylon normandii* Louvet, p. 98, pls 11–12, text-figs 23–25.

1971*b* *Entandrophragmoxylon normandii* Louvet, pp. 146, 148, pl. 3.

Discussion of woods from the Reading Formation

It is unusual to have many specimens of a species as most fossil woods are described from isolated specimens (Wheeler 1991; Wheeler *et al.* 1995). Although not found in growth position and fragmentary in nature, these fossils are therefore very useful, particularly for assessment of quantitative and qualitative variation (see Tables 6–7). The variation in the Reading woods is not as great as reported in *Paraphyllanthoxylon abbottae* by Wheeler (1991). Her quantitative range spanned most of the genera within *Paraphyllanthoxylon* and may be interspecific. The variation in the Reading samples, namely the tangential diameter of vessel lumina, vessel density and volume of ray parenchyma, is more likely to be a result of the relative position of the wood (i.e. branch, trunk, root). One of these remains is large; 1 m high and at least 70 cm in diameter (uncatalogued specimen in Newbury Museum). This would have been derived from a substantial tree. Growth rings are absent or indistinct in all the Reading woods, indicating no strong seasonal element to the climate; and the nearest Recent analogue for all of these woods, except *Fagoxylon*, is tropical.

A few of the woods have radial borings of two sizes, about 3 mm and 10 mm in diameter respectively, with no wall or contents. They are probably arthropod borings (Dr E. A. Jarzembowski, pers. comm. 1990) and possibly beetle borings (Dr J. Stephenson, pers. comm. 1990).

It is currently uncertain from which horizon the wood specimens came. Both marine (Crane and Goldring 1991) and freshwater (Reading) facies appear to be present at Chieveley.

Compression plant fossils also occur at Chieveley and appear similar to those from Cold Ash (Crane 1984), representing leaves and fructifications of *Cercidiphyllum*-like plants. These are not likely to have come from the woods described herein.

Woods from the Oldhaven Beds

Only one wood has been described from the Upper Pliocene conglomeratic lag phosphorite deposits at the base of the East Anglian Crags (Brett 1972). This is not surprising as the original provenance of the woods is doubtful. It was originally generally accepted that the woods were from the London Clay Formation (Brett 1972; Balson 1980). However a study of boring contents, preservation and anatomy of *Sabulia scottii* Stopes, 1912 from Herne Bay has confirmed that these fossils are from Palaeogene deposits, although probably the Oldhaven Beds rather than the London Clay. Specimens of phosphatized (apatite) wood are housed in the NHM and the Ipswich Museum form the basis of this study. Most represent small branch or root wood, usually with some primary wood remaining.

Relatively little is known about the plant remains of the Oldhaven Beds from fruit and seed remains (Chandler 1964). Further knowledge of this flora is especially important as it overlain by the London Clay Formation with its well documented and diverse floral assemblage.

The original provenance of Sabulia scottii *Stopes, 1912.* When Stopes described this wood in 1912, she stated that the specimen was from the Lower Greensand of Woburn in Bedfordshire (this is all the information known). She suggested that it may have been from a hill above the village of Woburn Sands. Here a bed of sand beneath Fuller's earth had yielded silicified gymnosperm wood. Another specimen, herein identified as *S. scottii* (V.17080), was donated in 1896 to the NHM as part of the Prestwich Collection. It is from Herne Bay, Kent and labelled Thanet Sand Formation. Its colour and preservation are identical to those of the holotype, indicating that it is from the same horizon (Pl. 5, figs 1–8). Examination of borings in this second specimen showed that some original matrix still remains. This is a poorly consolidated yellow-brown sandy clay with some shell debris, which, according to D. W. Ward, is identical to sediments of the Oldhaven Beds (uppermost Palaeocene/lowermost Eocene) at Herne Bay. Further study of the collections at the NHM revealed many more specimens of *S. scottii*. Stopes would probably not have been aware of these, which are in the Cainozoic collection, because she was employed solely to catalogue the Cretaceous collection at this time. They were all collected as derived fossils from the Phosphatic Nodule Bed (Plio-Pleistocene) of Suffolk, eastern England. As well as being the same species they again have exactly the same characteristic preservation and colour as the holotype. Usually they only differ from V.17080 in possessing a surface polish, probably caused by abrasion during reworking (P. Balson, pers. comm. 1989) and that the sedimentary fillings are now lithified. The holotype also has this polish, a feature typical of most animal and plant fossils from this bed, and importantly a critical feature that occurred post-fossilization. All the woods were previously thought to have originated from the London Clay Formation (Brett 1972; Balson 1980), probably because they are all phosphatized (apatite), but it now seems clear that these specimens are derived from the Oldhaven Beds of the Thames Basin. Stopes' *Sabulia* is one of these, and may have found its way to Bedfordshire as an erratic.

Another possible explanation is that most of the many specimens of *Sabulia* were in the collections when Stopes first published on the genus. They had all been re-registered at approximately the same time as the holotype as part of a large and poorly curated collection that was transferred from the Botany Department of the museum in January, 1898. Two Oldhaven woods from Suffolk are numbered V.5611 and V.5641, whereas the holotype is V.5654. Greensand gymnosperm wood was also part of this collection: V.5659 is a specimen of *Cupressinoxylon hortii* Stopes, 1915 from Woburn Sands, the stated locality of the holotype. Accidental translocation of information would seem, in my opinion, plausible.

Family ANNONACEAE Jussieu, 1789

Morphotaxon POLYALTHIOXYLON Bande, 1973

Type species. Polyalthioxylon parapaniense Bande, 1973, Palaeocene/Eocene Intertrappean Beds, India.

Polyalthioxylon oldhavenense sp. nov.

Text-figure 8

Derivation of name. After the Oldhaven Beds.

Holotype. NHM V.63248; slides V.63248$1–3.

Locality and horizon. Upper Pliocene conglomeratic lag phosphorite deposits at the base of the East Anglian Crags, Suffolk. Inferred to have been derived from uppermost Palaeocene/lowermost Eocene Oldhaven Beds.

Diagnosis. Growth rings indistinct. Vessels diffuse porous, mean density of 26 mm^2, solitary and in radial groups of 2–12, mostly 2–3, mean tangential diameter 99 μm, mean element length 325 μm, perforation plates simple, vessel to vessel pitting alternate, minute, bordered, 2–3 μm. Axial parenchyma banded, 8 mm^2, each band one cell wide. Mean ray density 7 mm^2, ray width 2–6 cells, mean ray height 707 μm or 29 cells, homocellular to heterogeneous, ray margins of 1–3 procumbent to square cells. Libriform fibres are present.

Description. An axis of 650 mm original diameter. No primary wood is preserved. Growth rings are present but not distinct [**2**]. The boundary is marked by longer radial vessel groups of up to 12 vessels, thick walled fibres and denser bands of parenchyma.

Vessel elements. Vessels diffuse porous [**5**] (Text-fig. 8A); solitary vessels are uncommon (5%) and in radial groups of 2–12 [**10**], usually 2–3 (Text-fig. 8C). Those that appear solitary have another much reduced vessel radially attached. Simple perforation plates [**13**]; intervessel pits alternate, minute [**24**], 2–3 μm; vessel to parenchyma pitting similar to intervessel [**30**]; some coalescent apertures. Tangential diameter of vessel lumina range 55–143 μm, mean 99 μm [**41**]. Density 17–45 mm^2, mean 26 mm^2 [**48**]. Vessel element length range 165–462 μm, mean 325 μm [**52**]; many of the vessels have a dark filling [**58**].

Imperforate tracheary elements. Libriform fibres usually thin walled but thick at growth ring boundary [**69**]; tangential diameter 12–25 μm.

Axial parenchyma. Uniseriate banded [**86**] and scalariform [**88**] (Text-fig. 8C); bands are 4–11 mm^2, mean 8 mm^2.

Ray parenchyma. Ray width 2–6 cells [**98**] (Text-fig. 8B); multiseriate ray height range 264–1210 μm (12–53 cells) [**102**], mean 707 μm (29 cells); ray width 33–88 μm, mean 52 μm (four cells). Composition is homocellular [**104**] to heterogeneous with 1–3 rows of procumbent to square marginal cells [**106–107**] (Text-fig. 8D). Density 5–10 mm^2, mean 7 mm^2 [**115**].

Comparisons. The main features of this fossil are vessels usually in radial groups; apotracheal parenchyma in numerous usually uniseriate bands; rays usually over 1 mm high and 2–8 cells wide. This is most similar to Recent Annonaceae. Comparison with Recent genera of Annonaceae is difficult owing to the uniform wood structure present in this family (Metcalfe 1987) but the structure of species of *Oxandra* and *Fusea* is generally close to that of the fossil.

Seven annonaceous fossil wood species have been described from Africa and Asia: *Annonoxylon edengense* Boureau, 1954, from the Eocene of l'Adrar Tiguirit, south Sahara; *A. striatum* Boureau, 1950*a*,

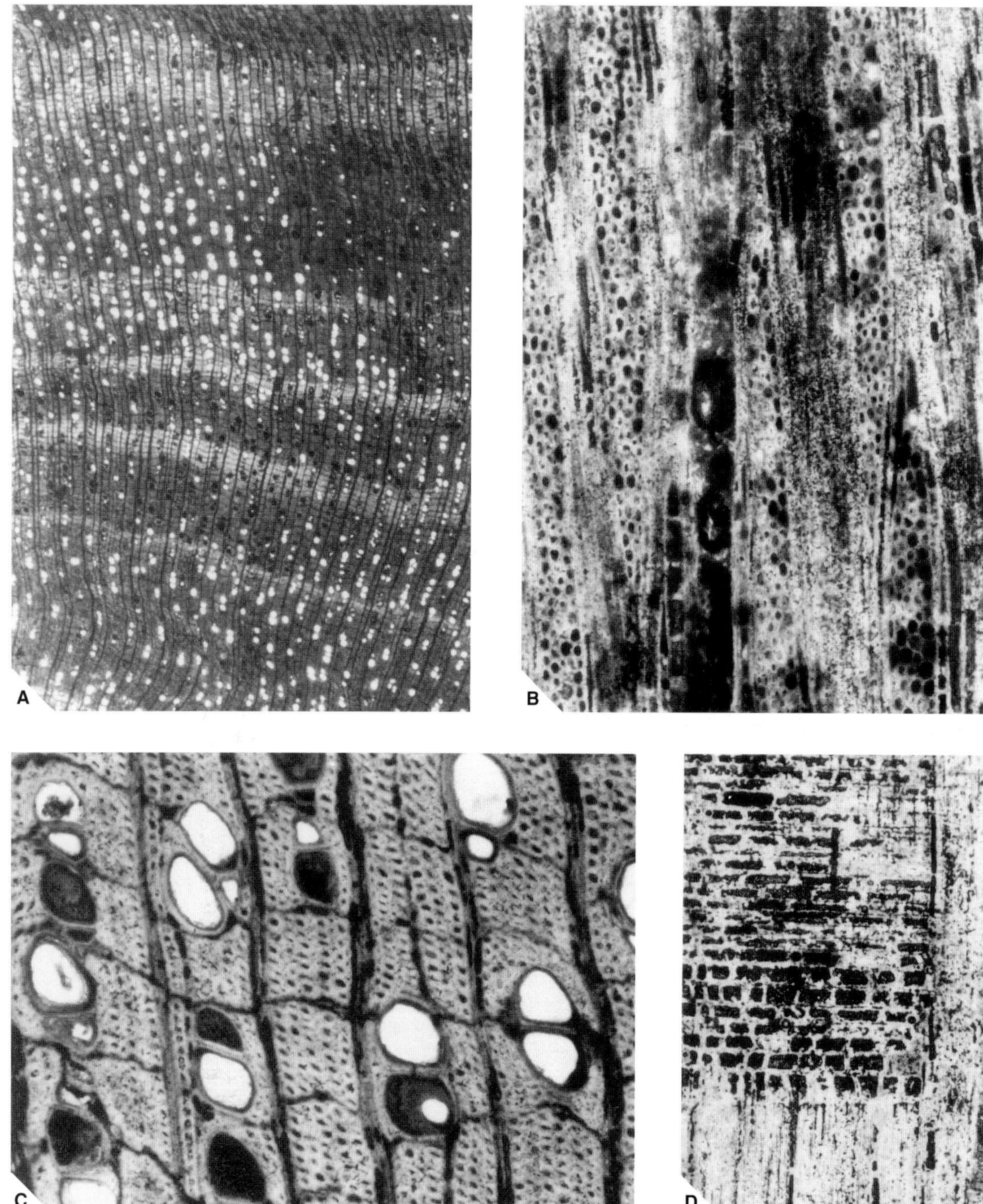

TEXT-FIG. 8. *Polyalthioxylon oldhavenense* sp. nov., Upper Pliocene conglomeratic lag phosphorite deposits at the base of the East Anglian Crags, Suffolk; inferred to have been derived from uppermost Palaeocene/lowermost Eocene Oldhaven Beds; holotype, NHM V.63248. A, TS, NHM V.63248$1; ×24. B, TLS, high multiseriate rays, NHM V.63248$2; ×240. C, TS, scalariform banded parenchyma, NHM V.63248$1; ×240. D, RLS, heterogeneous rays, NHM V.63248$3; ×240.

from the Tertiary? of Tamaguilel, south Sahara; *Polyalthia lateriflora* Kurz, 1874, recorded by Schuster 1911 from the Pliocene of Java; *Polyalthioxylon indicum* Prakash, 1978, from the middle Miocene Siwalik Beds of Uttar Pradesh, India; *P. parapaniense* Bande, 1973, from the Palaeocene or Eocene Deccan Intertrappean Beds of Parapani, Mandla District, India; *P. platymitroides* Kramer, 1974, and *P. stelechocarpoides* Kramer, 1974, from the Neogene of Java. All possess apotracheal banded parenchyma similar to that found in V.63248 except for *Polyalthia lateriflora*, which has diffuse-in-aggregates type. V.63248 is distinct from the other species by its minute pitting and longer radial vessel groups present at growth-ring boundaries. It is, therefore, described as a new species and included in *Polyalthioxylon* because it has non-septate fibres.

Remarks. No other specimens of Annonaceae are known from deposits older than the London Clay Formation. So far 14 species of seed have been described but have not been related to Recent genera (Collinson 1983). Today the family is chiefly tropical in its distribution (Collinson 1983).

Family APOCYNACEAE Jussieu, 1789, LECYTHIDCACEAE Poiteau, 1825 or SAPOTACEAE Jussieu, 1789

Morphotaxon APOCYNOXYLON Gazeau-Koeniguer, 1975

Type species. Apocynoxylon sylvestris Gazeau-Koeniguer, 1975, Eocene? (Lutetian?), Nucourt, France.

Apocynoxylon? *oldhavenense* sp. nov.

Text-figure 9

1989 Dicotyledons Pearson; p. 78 (V.8042).

Derivation of name. After the Oldhaven Beds.

Holotype. NHM V.63179; slides V.63179$1–3.

Other material. NHM V.63200, V.8042 (slide TS only).

Locality and horizon. Waldringfield, Suffolk (V.8042) and Suffolk (V.63179, V.63200); Upper Pliocene conglomeratic lag phosphorite deposits at the base of the East Anglian Crags; inferred to have been derived from uppermost Palaeocene/lowermost Eocene, Oldhaven Beds.

Diagnosis. Growth rings indistinct. Vessels diffuse porous, mean vessel density 13 mm^2, solitary and in radial multiples of 2–3, mean tangential diameter 91 μm, vessel to vessel pitting alternate, bordered, 2–4 μm. Axial parenchyma uniseriate banded, 5–15 mm^2, paratracheal scanty to vasicentric. Multiseriate ray height mean 1690 μm or 48 cells, range 336–3640 μm or 10–113 cells, heterogeneous with 1–6 marginal rows of square to upright cells. Libriform fibres present.

Description. Original diameter of axis at least 470 mm. Primary wood present but not well preserved. Growth rings present but indistinct [**2**], marked in part by a broad band of thick-walled fibres.

Vessel elements. Vessels diffuse porous [**5**] (Text-fig. 9A); solitary and in radial multiples of 2–4 (Text-fig. 9C). Simple perforation plates [**13**]. Vessel to vessel pits alternate, bordered, 2–4 μm [**24**]. Tangential diameter of vessel lumina range 55–132 μm, mean 91 μm [**41**]. Density 10–17 mm^2, mean 13 mm^2.

Imperforate tracheary elements. Libriform fibres [**66**] with medium-thick to thick walls [**69**], 10–20 μm in tangential diameter.

Axial parenchyma. Paratracheal scanty [**78**] to vasicentric [**79**]; banded (Text-fig. 9C), uniseriate [**86**], scalariform [**88**].

Ray parenchyma. Ray width 1–5 cells [**98**] (Text-fig. 9D); ray height, multiseriate range 336–3640 μm (10–113 cells), mean 1690 μm (48 cells) [**102**]; uniseriate height 260–784 μm (4–16 cells), mean 337 μm (nine cells). Composition is heterogeneous, marginal rows of usually 1–3 square to upright cells [**107**] (Text-fig. 9B) but sometimes up to eight cells [**108**]. Density 3–8 mm^2, mean 4–5 mm^2 [**115**].

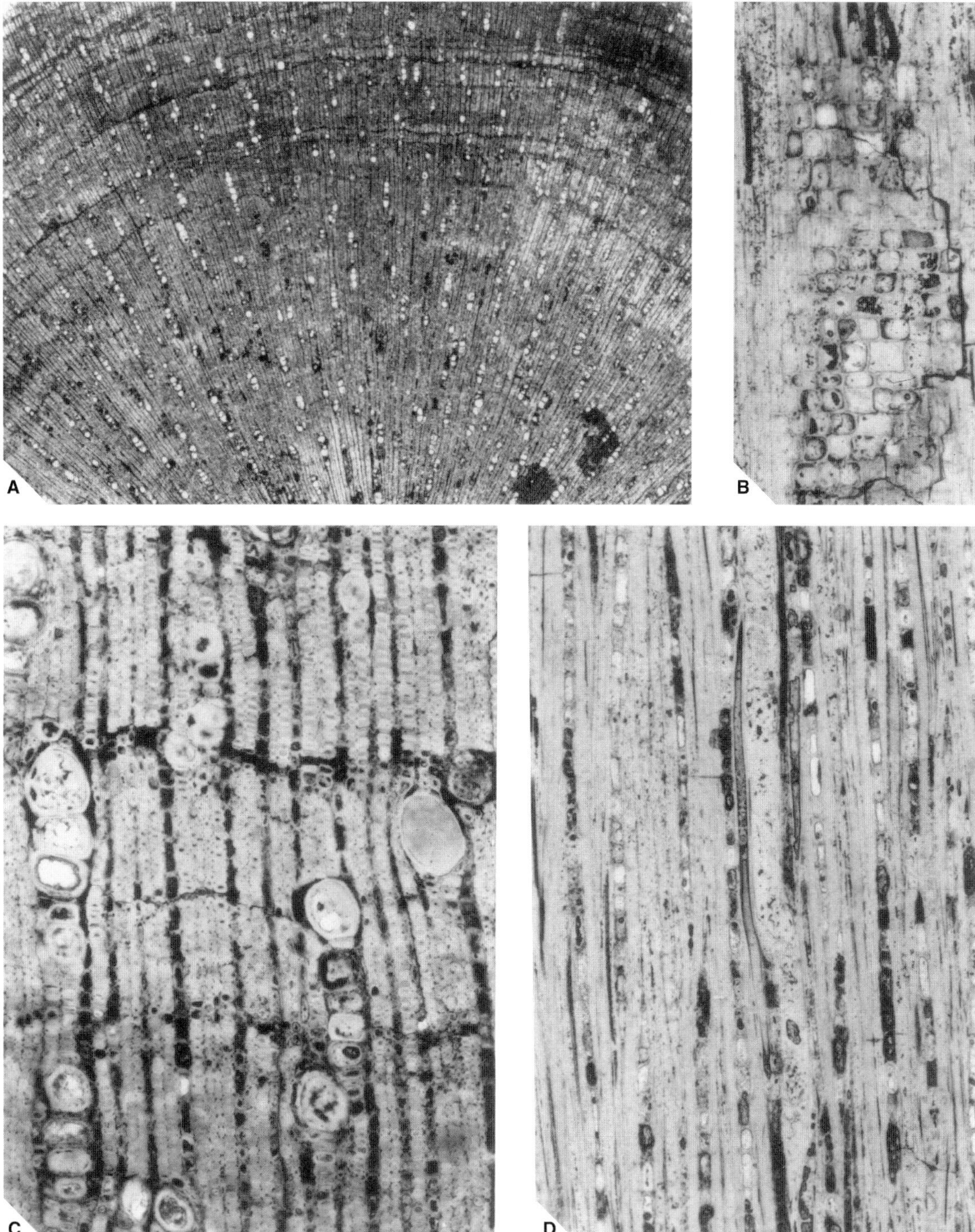

TEXT-FIG. 9. *Apocynoxylon*? *oldhavenense* sp. nov., Upper Pliocene conglomeratic lag phosphorite deposits at the base of the East Anglian Crags, Suffolk; inferred to have been derived from uppermost Palaeocene/lowest Eocene Oldhaven Beds; holotype, NHM V.63179. A, TS, NHM V.63179$1; ×24. B, TLS, rays, NHM V.63179$2; ×240. C, TS, banded and paratracheal parenchyma, NHM V.63179$1; ×240. D, RLS, heterogeneous ray cells, NHM V.63179$3; ×240.

Comparisons. The main features of this fossil wood are: abundant apotracheal banded parenchyma with some paratracheal parenchyma that is often scanty or vasicentric; and rays often more than 1 mm in height. In most respects this is similar to Recent woods of Apocynaceae and Lecythidaceae. There is also some local radial alignment of vessels in the fossil, which is a feature found in the Apocynaceae.

A single apocynaceous wood has been described, namely *Apocynoxylon sylvestris* Gazeau-Koeniguer, 1975, from the Eocene of France. Woods of lecythidaceous affinity are known from Cretaceous–Pleistocene deposits from south-east Asia and South America: *Barringtonioxylon arcotense* Awasthi, 1970, from the Upper Miocene–Pliocene Cuddalore succession of South India; *B. assamicum* Prakash and Tripathi, 1972, from the Upper Miocene Tipam succession of Assam, north-east India; *B. deccanense* Shallom, 1960, from the Lower Eocene Deccan Intertrappean Beds of Mahurzari, India; *B. eopterocarpum* Prakash and Dayal, 1965, from the Palaeocene–Eocene Deccan Intertrappean Beds of Mohgaon, Kalan, Keria, Mahurzari, and Nawargaon, India; *B. mandlaensis* Bande and Khatri, 1980, from the Palaeocene–Eocene Deccan Intertrappean Beds of Parapani, Mandla District, India; *Careyoxylon kuchilense* Prakash and Tripathi, 1972, from the Upper Miocene Tipam succession of Assam, north-east India; *C. pondicherriense* Awasthi, 1970, from the Upper Miocene–Pliocene Cuddalore succession of South India and the Neogene of Burma (Prakash and Bande 1980); *Lecythidoxylon brasiliense* Milanez, 1935, emend. Müller-Stoll and Mädel-Angeliewa, 1984, from the Cretaceous of Alegre Manga, Piaui, Brazil; *L. milanezii* Mussa, 1959, emend. Müller-Stoll and Mädel-Angeliewa, 1984, from the Tertiary? or Pleistocene? of Cachoeira do Gastao, Territorio do Acre, Brazil. Only the last two species have banded and paratracheal parenchyma similar to that found in V.63174. The Oldhaven wood differs by its much greater vessel density, uniseriate banded parenchyma, small intervascular pitting and ray height. Consequently it is regarded as a new species and is put into *Apocynoxylon* for wood of possible apocynaceous affinity.

Remarks (see *Apocynoxylon sapotaceoides* for comments on Apocynaceae). If lecythidiaceous, this would be the first report of this family from Britain. Lecythidiaceae are pantropical; the New World woods have crystalliferous parenchyma unlike their Old World counterparts (and the Oldhaven wood).

Apocynoxylon sapotaceoides sp. nov.

Text-figure 10

Derivation of name. From a similarity to woods in the family Sapotaceae.

Holotype. IM 2; slides IM 2$1–3.

Locality and horizon. Unlocalised Upper Pliocene conglomeratic lag phosphorite deposits at the base of the East Anglian Crags, Suffolk; inferred to have been derived from uppermost Palaeocene/lowermost Eocene, Oldhaven Beds.

Diagnosis. Growth rings indistinct. Vessel mean density 20 mm^2. Vessels arranged in a distinct radial pattern, solitary vessels 5%, radial groups of 2–12, vessel to vessel pitting alternate, bordered and minute, 3–4 μm. Axial parenchyma banded, up to three cells wide. Rays exclusively uniseriate, mean height 887 μm or 18 cells, mean density 15 mm^2, heterogeneous. Libriform fibres are present.

Description. Growth ring boundaries indistinct [**2**], marked by flattened fibres and axial parenchyma?

Vessel elements. Vessels diffuse porous [**5**], in radial pattern [**7**] (Text-fig. 10A); solitary vessels 5% and in radial groups of 2–12; 70% are in groups of 4–9 [**10**]. Simple perforation plates [**13**]. Intervessel pits alternate [**22**], bordered; pit size minute [**24**], 3–4 μm; vessel to parenchyma pits not seen. Tangential diameter of vessel lumina range 33–110 μm, mean 81 μm [**41**]. Density 14–28 mm^2, mean 20 mm^2 [**47–48**].

Imperforate tracheary elements. Libriform fibres, no pitting seen [**61**], thin to thick walled [**69**].

Axial parenchyma. Banded parenchyma, narrow bands up to three cells wide [**86**] (Text-fig. 10C).

Ray parenchyma. Exclusively uniseriate rays [**96**] (Text-fig. 10D); height range 187–1870 μm (4–42 cells), mean 887 μm (18 cells). Composition is heterogeneous (Text-fig. 10B). Density 11–19 rays mm^2, mean 15 mm^2 [**116**].

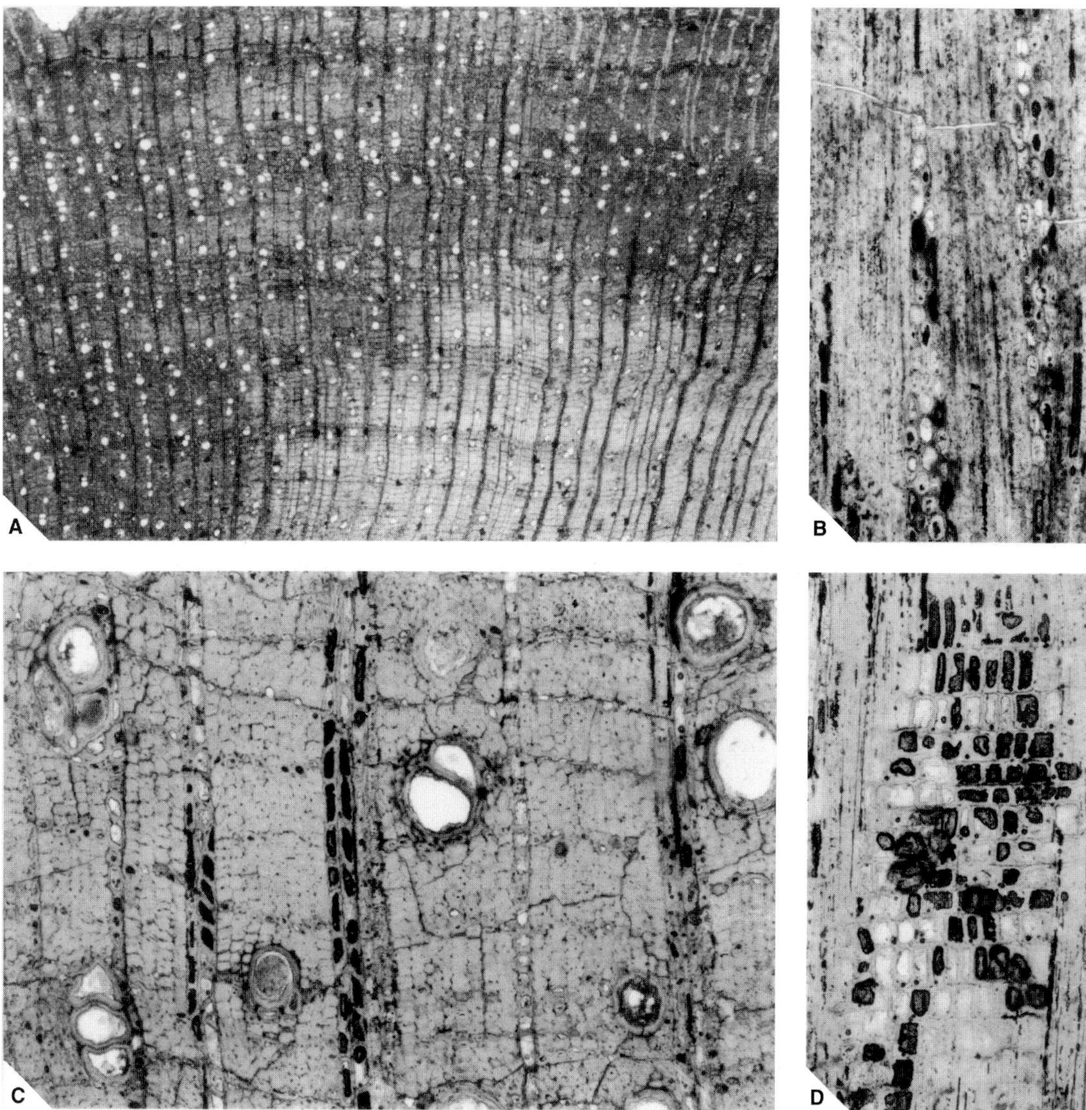

TEXT-FIG. 10. *Apocynoxylon sapotaceoides* sp. nov., unlocalised Upper Pliocene conglomeratic lag phosphorite deposits at the base of the East Anglian Crags, Suffolk; inferred to have been derived from uppermost Palaeocene/ lowermost Eocene Oldhaven Beds, holotype, IM 2. A, TS, note radial alignment of vessel groups, IM 2$1; ×24. B, RLS, heterogeneous rays, IM 2$3; ×240. C, TS, showing banded parenchyma, IM 2$1; ×240. D, TLS, exclusively uniseriate rays, IM 2$2; ×240.

Comparisons. The main features of the fossil are: vessels in a radial pattern and predominantly in long radial groups; banded parenchyma; exclusively uniseriate and heterogeneous medullary rays. This combination is found in Recent Sapotaceae and Apocynaceae. Of Sapotaceae, the structure is very close to the genus *Neoxythece*, particularly *N. cladantha* (Sandwith) Aubréville, 1961 (Detiénne and Jacquet 1983) from the Amazon Basin. Three genera of Apocynaceae, *Cameraria, Plumeria* and *Zschokkea*, have species with uniseriate banded parenchyma and exclusively uniseriate rays. Woods of this family usually lack vasicentric tracheids unlike those of Sapotaceae in which they are present in most genera. Vasicentric tracheids were not observed in the fossil so it seems more likely that it is apocynaceous.

Fossil woods similar to the Suffolk fossil are *Apocynoxylon sylvestris* Gazeau-Koeniguer, 1975 (Sapotaceae or Apocynaceae?) and *Ebenoxylon aegyptiacum* Kräusel, 1939 (Ebenaceae). *Ebenoxylon aegyptiacum* has a higher percentage of solitary vessels of between 40–50% whilst *Apocynoxylon sylvestris* has rays 1–4 cells wide and parenchyma bands up to seven cells wide. Hence, the Oldhaven woods are described as *Apocynoxylon sapotaceoides* sp. nov., which is intended for woods with a structural pattern seen in Recent *Neoxythece* (Sapotaceae), *Cameria, Plumeria* and *Zschokkea* (Apocynaceae).

Remarks. No other apocynaceous or sapotaceous remains are known from the Oldhaven Beds but possible representatives of the Apocynaceae occur in the underlying Woolwich and Reading Beds, and are recorded from the overlying London Clay Formation (Chandler 1964). Extant members of this family are tropical to subtropical in distribution. Sapotaceous remains are known from the London Clay (Chandler 1964); the current distribution of this family is pantropical.

Family CAESALPINACEAE? R. Brown, *in* Flinders, 1814, or FABACEAE? Lindley, 1836

Morphotaxon TETRAPLEUROXYLON Müller-Stoll and Mädel, 1967

Type species. Tetrapleuroxylon ingaeforme (Felix) Müller-Stoll and Mädel, 1967, Brazil.

Tetrapleuroxylon oldhavenense sp. nov.

Text-figure 11

Derivation of name. After the Oldhaven Beds.

Holotype. NHM V.63175; slides V.63175$1–3.

Other material. NHM V.63201.

Locality and horizon. Unlocalised Upper Pliocene conglomeratic lag phosphorite deposits at the base of the East Anglian Crags, Suffolk; inferred to have been derived from uppermost Palaeocene/lowermost Eocene, Oldhaven Beds.

Diagnosis. Growth rings distinct to vague, marked by small vessels and broken marginal parenchyma. Vessels diffuse porous, solitary and radial groups of 2–5, perforation plates simple, mean tangential diameter 127 μm, vessel to vessel pitting alternate, bordered, 4–7 μm. Parenchyma paratracheal, vasicentric to lozenge aliform and confluent, connecting up to eight vessel groups obliquely to tangentially and forming broken, wavy bands; diffuse parenchyma present, scarce. Rays 1–4 cells wide, usually three cells, mean height 383 μm or 24 cells, heterogeneous with one row of square marginal cells; uniseriate rays very scarce. Crystals in chambered parenchyma cells.

Description. No primary wood is present. The growth rings are indistinct [**2**], marked by very small vessels and broken marginal parenchyma with boundaries 6·5–10 mm apart.

Vessel elements. Vessels diffuse porous [**5**], solitary and in radial groups of 2–5; perforation plates simple [**13**]. Vessel to vessel pits alternate, bordered, small 4–6 μm [**25**]; vessel to parenchyma pitting similar to vessel to vessel [**30**]. Tangential diameter of vessel lumina range 55–187 μm, mean 127 μm [**42**]. Density 6–15 mm^2, mean 9–10 mm^2 [**47**]. Vessel element length range 110–418 μm, mean 254 μm [**52**].

Imperforate tracheary elements. Libriform fibres [**66**]; tangential diameter 10–20 μm, medium-thick to thick walls [**69**].

Axial parenchyma. Abundant, paratracheal, vasicentric [**79**] to lozenge aliform [**81**] (Text-fig. 11A, bottom), sheaths up to eight cells broad, confluent [**83**], connecting up to eight vessel groups obliquely or tangentially (Text-fig. 11C);

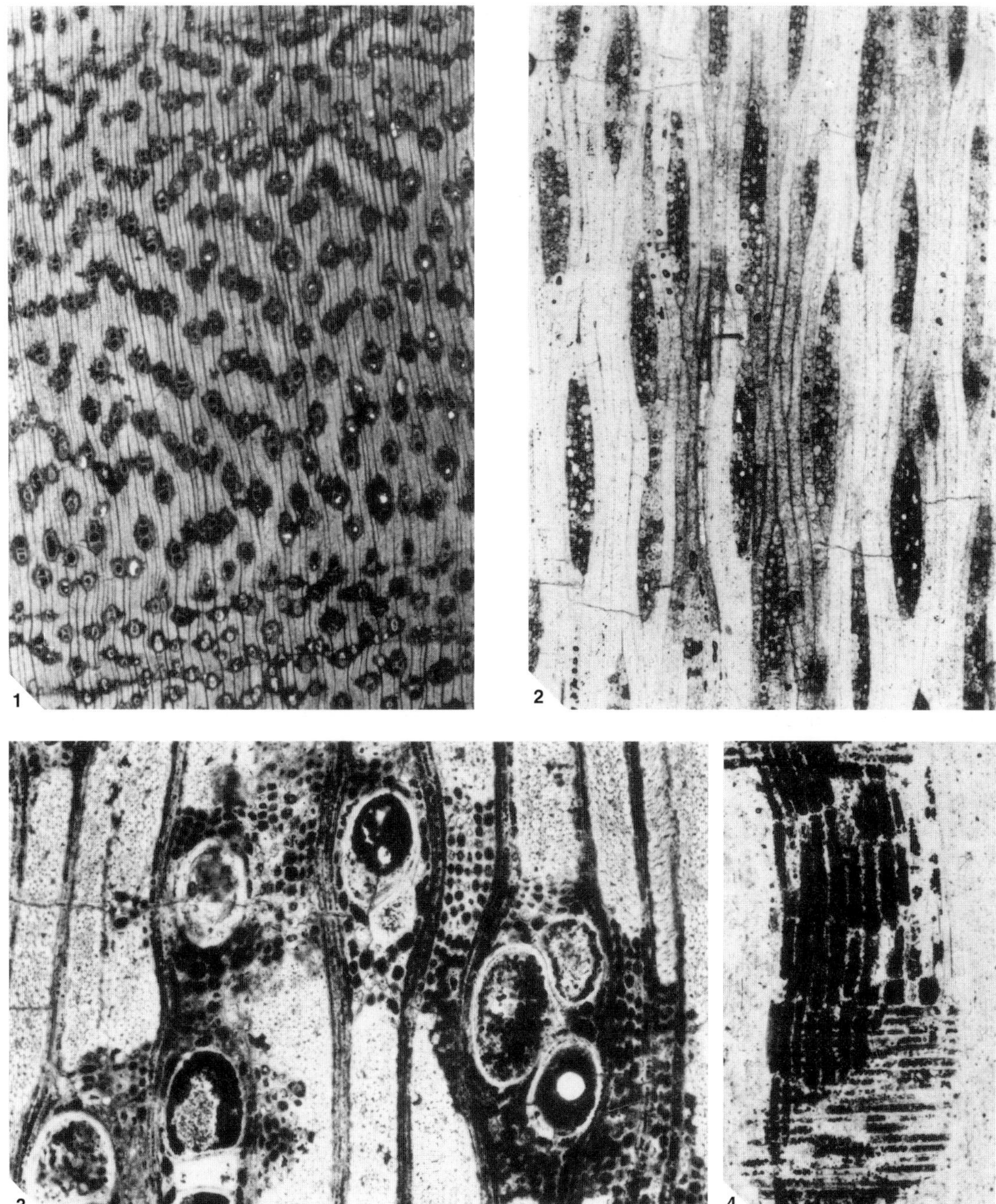

TEXT-FIG. 11. *Tetrapleuroxylon oldhavenense* sp. nov., unlocalised Upper Pliocene conglomeratic lag phosphorite deposits at the base of the East Anglian Crags, Suffolk; inferred to have been derived from uppermost Palaeocene/lowermost Eocene Oldhaven Beds, holotype, NHM V.63175. A, TS, vague growth ring can be seen towards the bottom of this figure, NHM V.63175$1; ×24. B, TLS, chambered crystalliferous parenchyma can be seen near the centre of this figure, NHM V.63175$2; ×240. C, TS, paratracheal confluent parenchyma, NHM V.63175$1; ×240. D, RLS, heterogeneous rays and storied axial parenchyma, NHM V.63175$3; ×240.

also diffuse as rare, isolated strands distinguished from the fibres in TS by their thin walls and brown-coloured contents. Storied structure present [**120**] (Text-fig. 11D).

Ray parenchyma. Ray width 1–4 cells [**98**] (Text-fig. 11B); multiseriate ray height range 110–726 μm (7–51 cells), mean 383 μm (24 cells); uniseriate ray height range 1–9 cells, mean five cells. Composition is heterogeneous, a single row of upright marginal cells [**106**] (Text-fig. 11D). Density 7–12 mm^2, mean 8–9 mm^2 [**115**].

Mineral inclusions. Prismatic crystals in chambered axial parenchyma cells [**142**] (Text-fig. 11B, bottom left), containing up to 23 crystals.

Comparisons. The overall structure of this specimen, and in particular the parenchyma, pitting and ray structure, is similar to that of woods of the Caesalpinaceae or Fabaceae. Leguminous fossil woods with parenchyma that is predominantly paratracheal vasicentric to confluent, diffuse and marginal with narrow rays, are described under the genus *Tetrapleuroxylon* Müller-Stoll and Mädel, 1967. The following species have been described: *Tetrapleuroxylon acaciae* (Kräusel) Müller-Stoll and Mädel, 1967, from the Tertiary of Egypt and the Palaeogene of the Grand Erg Oriental, Algeria (Louvet 1971*a*); *T.* aff. *acaciae* (Kräusel) Müller-Stoll and Mädel, 1967, from the Oligocene of Tunisia (Delteil-Desneux and Fessler-Vrolant 1976); *T. aquitanense* Dupéron, 1975, from the Oligocene (Stampian) of Agenais, France; *T. communis* Petrescu, 1978, from the Oligocene of Romania; *T. ersanense* (Boureau) Müller-Stoll and Mädel, 1967, from the Eocene of the Republic of Mali and the Tertiary of Québrada de Culluhuay, Lima, Peru (Salard 1963); *T. ingaeforme* (Felix) Müller-Stoll and Mädel, 1967, of unknown age from Brazil; *T. limagnense* Privé, 1973, from the Pliocene of L'Allier, France; T. vantagiensis (Prakash and Barghoorn) Müller-Stoll and Mädel, 1967, from the Miocene Columbia basalts of Washington, USA; and *T. zaccarinii* (Chiarugi) Müller-Stoll and Mädel, 1967, from the Upper Cretaceous of Somalia. It is closest to *Tetrapleuroxylon acaciae* (Louvet 1971*a*), *T.* aff. *acaciae* and *T. vantagiensis* but can be distinguished by the combination of wavy banded parenchyma, scarce diffuse parenchyma and heterogeneous rays, and rare uniseriate rays. However, as pointed out by Wheeler and Baas (1992), care should be taken when assigning fossil wood to the Leguminosae because several other families have similar features, e.g. the Bignoniaceae, Sapindaceae and Rutaceae. The presence of vestured pitting in legumes is often critical in making such a distinction and may be determined in fossil material (Crawley 1988). It was not observed in the Oldhaven fossil; hence the family assignment for this fossil is tentative.

Remarks. No other leguminous species have been recorded from the Oldhaven Beds, but two species are known from the older Reading Formation and Branksome Sand (late Early–Mid Eocene) (Chandler 1964; Herendeen and Crane 1992). Both are considered to be referable to the Caesalpiniaceae.

Family CERCIDIPHYLLACEAE Engler, 1909, HAMAMELIDACEAE R. Brown, *in* Tuckey, 1818, or ICACINACEAE (Bentham) Miers, 1851

Morphotaxon CERCIDIPHYLLOXYLON Prakash, Brezinova and Bužek, 1971

Type species. Cercidiphylloxylon kadanense Prakash, Brezinova and Bužek, 1971, Oligocene?, Czechoslovakia.

EXPLANATION OF PLATE 11

1–5. *Cercidiphylloxylon spenceri* (Brett) Pearson, 1987, unlocalised, Upper Pliocene conglomeratic lag phosphorite deposits at the base of the East Anglian Crags, Suffolk; inferred to have been derived from uppermost Palaeocene/lowermost Eocene Oldhaven Beds, NHM V.5008. 1, TS, solitary vessels, NHM V.5008$1; ×240. 2, RLS, scalariform perforation plate, NHM V.5008$3; ×960. 3, TS, NHM V.5008$1; ×24. 4, RLS, heterogeneous rays, NHM V.5008$3; ×240. 5, TLS, rays with long uniseriate 'tails', NHM V.5008$2; ×240.

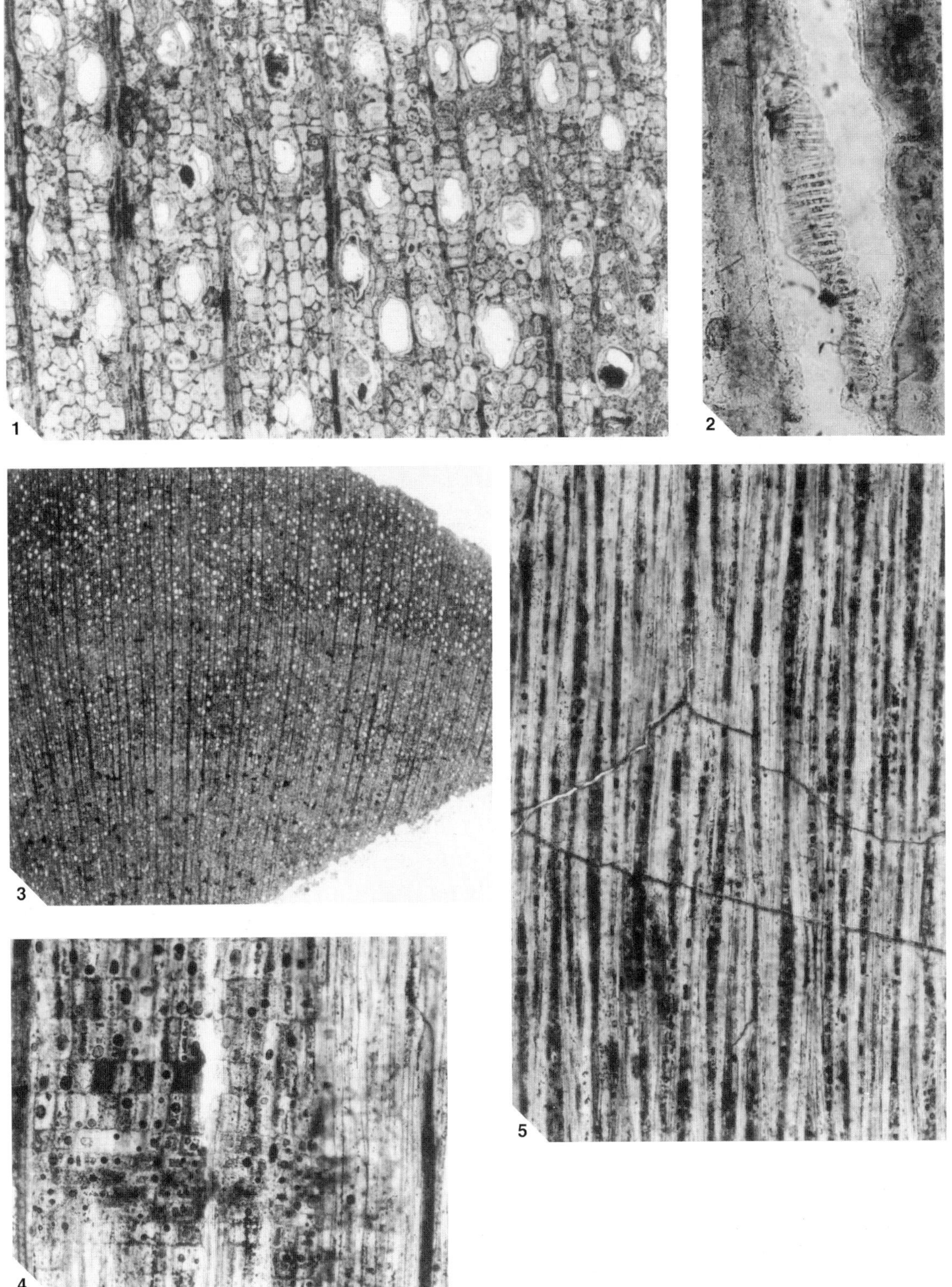

CRAWLEY, *Cercidiphylloxylon*

Cercidiphylloxylon cf. *spenceri* (Brett) Pearson, 1987

Plate 11

1989 Dicotyledons Pearson; p. 78 (V.8010 and V.8032).

Material. NHM V.5008; slides V.5008$1–3, V.8010 and V.8032 (both TS slides only).

Locality and horizon. Waldringfield, Suffolk (V.8010 and V.8032), and unlocalised Suffolk (V.5008); Upper Pliocene conglomeratic lag phosphorite deposits at the base of the East Anglian Crags; inferred to have been derived from uppermost Palaeocene/lowermost Eocene Oldhaven Beds.

Description. An axis 4 cm in diameter. No primary wood preserved. Growth rings are present and distinct (Pl. 11, fig. 3) [**1**], marked by rows of flattened fibre-tracheids with boundaries 3–5 mm apart.

Vessel elements. Vessels diffuse porous [**5**], almost entirely solitary (Plate 11, fig. 1) [**9**]; apparent tangential pairs are seen where vessel element ends overlap. Perforation plates are exclusively scalariform (Pl. 11, fig. 2) [**14**], at an oblique angle, with 24–36 bars, mean 29 bars [**17**]. Vessel to vessel pitting not seen, probably owing to solitary vessels; vessel to parenchyma pitting opposite to transitional [**31–32**], with pit apertures 4–6 μm in diameter. Tangential diameter of vessel lumina range 28–68 μm, mean 52 μm [**41**]. Density of vessels 18–54 mm^2, mean 30 mm^2 [**48**].

Imperforate tracheary elements. Fibre-tracheids; tangential diameter 22–25 μm; distinctly bordered pits [**62**]; thick walls (Pl. 11, fig. 1) [**70**].

Axial parenchyma. Apotracheal diffuse and diffuse-in-aggregates (Pl. 11, fig. 1) [**76–77**].

Ray parenchyma. Ray width 1–4 cells (Pl. 11, fig. 5) [**98**] (most of the following cell counts are estimates a result of poor preservation): multiseriate ray height range 228–1320 μm (8–36 cells), mean 635 μm (22 cells); uniseriate ray height range 222–770 μm (5–16 cells), mean 546 μm (10–11 cells); ray width range 11–77 μm (1–4 cells), mean 34 μm (two cells). Composition is heterogeneous; ray margins composed of more than four rows of upright cells (Pl. 11, fig. 4) [**108**]. Density is 12–20 mm^2, mean 15–16 mm^2 [**116**].

Comparisons. The main features of V.5008 are: vessels with transitional to opposite intervascular pitting and scalariform perforation plates, apotracheal diffuse and diffuse-in-aggregates parenchyma, narrow and heterogeneous rays and fibre-tracheids. This structure is possessed by eight Recent genera in seven families; *Apodytes* (Icacinaceae); *Cercidiphyllum* (Cercidiphyllaceae); *Curtisia* (Cornaceae); *Eurya* (Theaceae); *Liquidambar* and *Rhodoleia* (Hamamelidaceae); *Styloceras* (Buxaceae); and *Weinmannia* (Cunoniaceae). Of these, probably *Curtisia*, *Apodytes* and *Eurya* are structurally the most similar to the fossil described here.

Seven Tertiary woods with a structure similar to that of V.5008 have been described: *Cercidiphylloxylon kadanense* Prakash, Brezinova and Bužek, 1971 (Cercidiphyllaceae), from the Oligocene? of the Czech Republic/Slovakia; *C. spenceri* (Brett) Pearson, 1989, from the Palaeocene Interbasaltic Beds of the Plateau Volcanics Series (Crawley 1989), the Eocene Clarno Formation (Scott and Wheeler 1982, as *C. alalongum*), and the London Clay (Brett 1956); *Hamamelidoxylon castellanense* Grambast-Fessard, 1969, from the Upper Miocene of France; *H. rhenanum* Burgh, 1973, from the Miocene of Germany; *Liquidambar speciosum* Felix, 1884 (Burgh 1964), also from the Miocene of Germany (all Hamamelidaceae). The possession of commonly three biseriate portions per ray, with the uniseriate portions of the ray often no wider than the multiseriate portions, show the Oldhaven wood is most similar to *Cercidiphylloxylon*, particularly *C. spenceri*, although the latter has narrower rays. This could be intraspecific variation; as a result the specimens examined are referred here to *Cercidiphylloxylon* cf. *spenceri*.

Remarks. A single species of icacinaceous seed is recorded from the Oldhaven Beds whereas other icacinaceous, hamamelidaceous and cercidiphyllidaceous remains are known from the Mull Interbasaltic Beds, Woolwich and Reading Formation, and the London Clay Formation.

Family FLACOURTIACEAE Richard ex de Candolle, 1824

Morphotaxon FLACOURTIOXYLON Louvet, 1970

Type species. Flacourtioxylon gifaense Louvet, 1970, Eocene (Upper Lutetian), Libya.

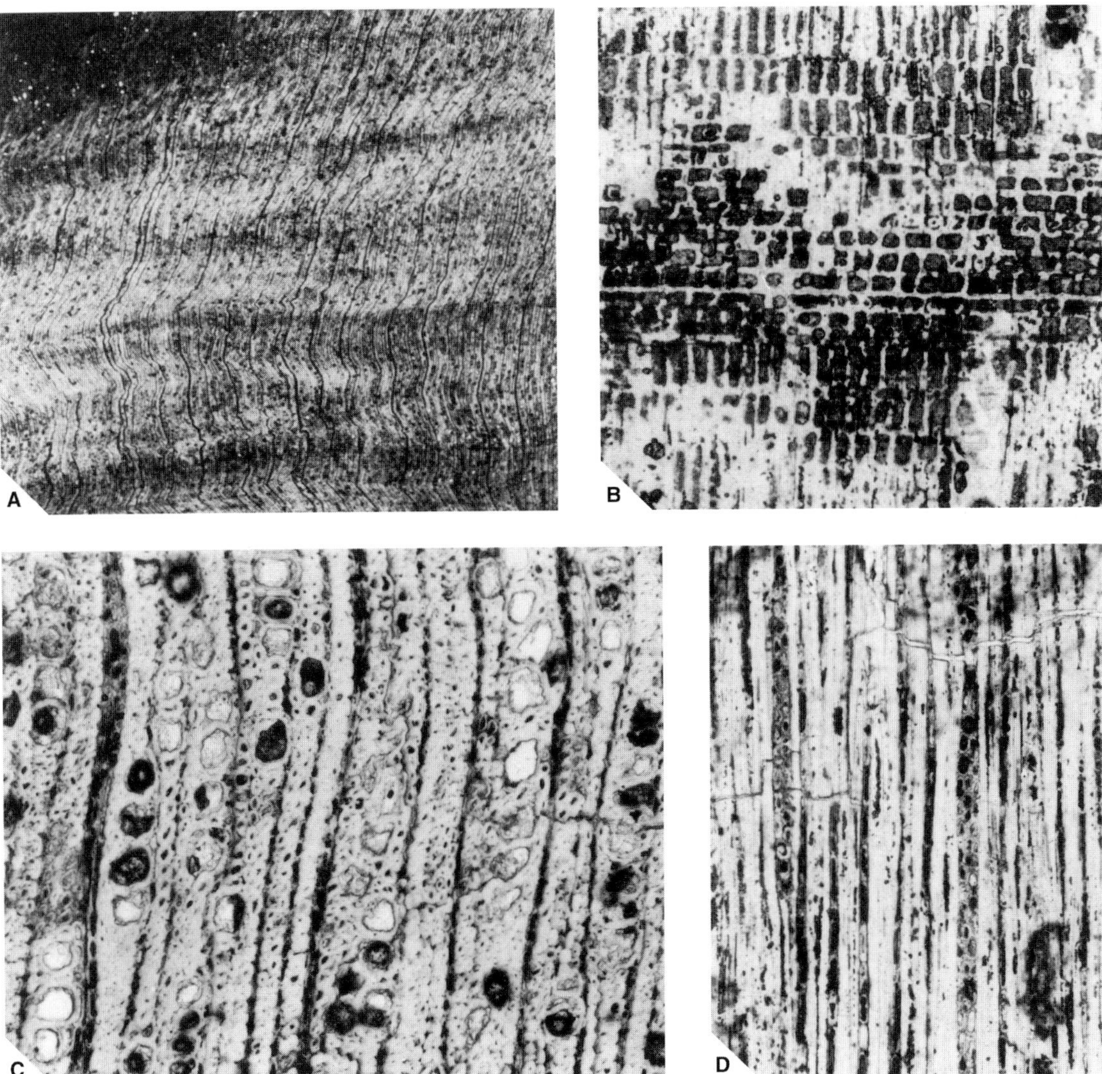

TEXT-FIG. 12. *Flacourtioxylon oldhavenense* sp. nov., Upper Pliocene conglomeratic lag phosphorite deposits at the base of the East Anglian Crags, Suffolk; inferred to have been derived from uppermost Palaeocene/lowermost Eocene Oldhaven Beds, holotype, NHM V.63173. A, TS, NHM V.63173$1; ×24. B. RLS, heterogeneous rays with several rows of upright marginal cells, NHM V.63173$3; ×240. C, TS, vessels in radial groups and lack of axial parenchyma, NHM V.63173$1; ×240. D, TLS, multiseriate and uniseriate rays, NHM V.63173$2; ×240.

Flacourtioxylon oldhavenense sp. nov.

Text-figure 12

1989 Dicotyledons Pearson; p. 78 (V.8016).

Derivation of name. After the Oldhaven Beds.

Holotype. NHM V.63173; slides V.63173$1–3.

Other material. NHM V.8016 (slide TS only).

Locality and horizon. Waldringfield, Suffolk (V.8016); others unlocalised Suffolk; Upper Pliocene conglomeratic lag phosphorite deposits at the base of the East Anglian Crags; inferred to have been derived from uppermost Palaeocene/ lowermost Eocene, Oldhaven Beds.

Diagnosis. Growth rings indistinct. Diffuse porous, mean density 38 mm^2, vessels solitary and in radial groups of 2–9, often with radial alignment of groups; mean tangential diameter 38 μm, vessel to vessel pitting alternate, bordered and 5–6 μm in horizontal diameter. Axial parenchyma absent or extremely rare. Mean ray density 20 mm^2, mean multiseriate ray height 787 μm, mean uniseriate ray height 602 μm, uniseriate rays abundant, multiseriate rays heterogeneous with ray margins of up to eight upright cells. Imperforate tracheary elements, libriform non-septate fibres. Solitary prismatic crystals present in procumbent ray cells.

Description. Axis originally at least 4 cm in diameter. No primary wood is present. Growth rings are present but indistinct [**2**], marked by longer vessel chains.

Vessel elements. Vessels diffuse porous [**5**]; some radial pattern [**7**] (Text-fig. 12A); solitary vessels and in radial groups of 2–4 [**10**] (Text-fig. 12C), locally 5–9; simple perforation plates [**13**]. Vessel to vessel pits alternate, bordered [**22**], 5–6 μm [**25**]; vessel to parenchyma pits not seen. Tangential diameter of vessel lumina range 22–63 μm, mean 46 μm [**40**]. Density 20–60 mm^2, mean 38 mm^2 [**48**]. Some of the vessels are filled with a black substance [**58**] and tyloses are present, some possibly sclerotic [**57**].

Imperforate tracheary elements. The fibres are septate in part [**65, 66**] and thin to thick walled [**69**]; tangential diameter 10–15 μm.

Axial parenchyma. No parenchyma was observed in TS although some possible strands were seen in TLS [**75**].

Ray parenchyma. Ray width 1–4 cells [**98**]; multiseriate ray height range 330–1375 μm (15–49 cells) [**102**], mean 787 μm (30 cells); multiseriate ray width range 15–66 μm (2–4 cells), mean 29 μm (two cells) (Text-fig. 12D). Composition is heterogeneous with ray margins of up to eight upright cells [**108**] (Text-fig. 12B). Density 14–26 mm^2, mean 20 mm^2 [**116**].

Mineral inclusions. Solitary prismatic crystals are present in ordinary procumbent ray cells [**138**].

Comparisons. This wood has a structure that is typical of Flacourtiaceae; vessels that are solitary and commonly in radial multiples of 2–4; parenchyma either absent or rare; rays that are markedly heterogeneous, usually with margins of 4–10 cells in species in which the ray height is under 1 mm; septate fibres; and intervascular and vessel to ray pitting that is very small to minute. Comparison with Recent genera of similar structure, such as *Homalium*, *Idesia* and *Xylosomia*, show a general resemblance to the fossil.

Seven fossil wood species of flacourtiaceous affinity have been described: *Aphloioxylon groenlandicum* Mathiesen, 1961, from the Palaeocene (Danian) of Greenland; *Casearioxylon tibatiense* Dupéron-Laudoueneix, 1977, from the Cretaceous? of the River Djerem, Mbakaou, Tibati, Cameroon; *Flacourtioxylon*? sp. Greguss, 1969, from the middle Eocene of Tatabanya, Hungary; *F.* (*Monimiaoxylon*) *gifaense* Louvet, 1970, from the Eocene (Lutetian) of Libya; *Homalioxylon assamicum* Prakash and Tripathi, 1974, from the Upper Miocene Tipam succession of Assam, north-east India; *H. mandlaense* Bande, 1974, and *Hydnocarpoxylon indicum* Bande and Khatri, 1980, from the Palaeocene–Eocene Deccan Intertrappean Beds of Parapani, Mandla District, India. Only two of these species bear close comparison with V.63173, namely *Casearioxylon tibatiense* and *Flacourtioxylon gifaense*. V.63173 is distinct from both particularly in having radial groups of 4+, radial alignment of the vessel groups, and narrower rays with higher ray density. It is, therefore, regarded as a new species of *Flacourtioxylon*.

Remarks. No other flacourtiaceous remains are recorded from the Oldhaven Beds but a possible fruit has been reported from the older Woolwich and Reading Formation (Chandler 1964). Three genera of fruits are known from the overlying London Clay Formation (Collinson 1983). Today this family has a mainly tropical and subtropical distribution (Collinson 1983).

Family HAMAMELIDACEAE R. Brown, *in* Tuckey, 1818

Morphotaxon HAMAMELIDOXYLON Lignier, 1907

TEXT-FIG. 13. *Hamamelidoxylon renaultii* Lignier, 1907, Upper Pliocene conglomeratic lag phosphorite deposits at the base of the East Anglian Crags, Suffolk; inferred to have been derived from uppermost Palaeocene/lowermost Eocene Oldhaven Beds, NHM V.63176. A, TS, distinct growth rings, zones of varying widths, NHM V.63176$1; ×24. B, TLS, exclusively uniseriate rays, NHM V.63176$2; ×240. C, TS, exclusively solitary vessels, NHM V.63176$1; ×240. D, RLS, heterogeneous rays, NHM V.63176$3; ×240. E, RLS, scalariform perforation plate, NHM V.63176$3; ×600.

Type species. Hamamelidoxylon renaultii Lignier, 1907, Upper Cretaceous (Cenomanian), France.

Hamamelidoxylon renaultii Lignier, 1907

Text-figure 13

1907 *Hamamelidoxylon renaultii* Lignier, pp. 300–301, pl. 19, figs 44–52; pl. 20, fig. 68; pl. 23, figs 85, 93; text-figs 3–5.
1969 *Liquidambaroxylon maegdefrauii* Greguss, p. 45, pl. 25.
1973 *Hamamelidoxylon maegdefrauii* (Greguss) Burgh, p. 180.

Material. NHM V.63176; slides V.63176$1–3.

Locality and horizon. Unlocalised Upper Pliocene conglomeratic lag phosphorite deposits at the base of the East Anglian Crags, Suffolk; inferred to have been derived from uppermost Palaeocene/lowermost Eocene Oldhaven Beds.

Description. Axis originally at least 9 cm in diameter. No primary wood is preserved. Growth rings are distinct [**1**] and marked by a zone of radially flattened fibre-tracheids with boundaries 2–7 cm apart (Text-fig. 13A).

Vessel elements. Vessels diffuse porous [**5**] (Text-fig. 13C), exclusively solitary [**9**], angular in outline [**12**]. Exclusively scalariform perforation plates [**14**] with at least 32 bars [**17**] (Text-fig. 13E); pitting not observed. Tangential diameter of vessel lumina range 22–46 μm, mean 36 μm [**40**]. Density 119–200 mm^2, mean 173 mm^2. Vessel element length not observed.

Imperforate tracheary elements. Fibre-tracheids with distinctly bordered pits [**62**], 10–30 μm in tangential diameter and of indeterminate length; medium-thick to thick walls [**69**].

Axial parenchyma. Apotracheal diffuse [**76**].

Ray parenchyma. Exclusively uniseriate [**96**] (Text-fig. 13B) except for a single biseriate ray; ray height range 110–792 μm (5–35 cells), mean 451 μm (22 cells). Composition is heterogeneous, composed of procumbent, square and upright cells [**109**] (Text-fig. 13D); density 13–23 mm^2, mean 18 mm^2 [**116**].

Comparisons. This structure is typical of the wood of *Exbucklandia, Fothergilla*, *Hamamelis* and *Parrotiopsis* (all Hamamelidaceae). When closely compared to the fossil, *Exbucklandia* has larger vessels whilst the other three have under 20 bars in their perforation plates.

Six fossil woods have been described with this characteristic structure: *Apodytoxylon hamamelidoides* Grambast-Fessard, 1969, from the Upper Miocene of France; *Corylopsites groenlandicus* Mathiesen, 1932, of unknown age from Greenland; *Hamamelidoxylon maegdefrauii* (Greguss) Burgh, 1973, from the Miocene (Helvetian) of Szarvasko, Hungary; *H. renaultii* Lignier, 1907, from the Cretaceous of Normandy, France; *H. rhenanum* Burgh, 1973, from the Miocene Brown Coal of the north Rhineland region of Germany; and *H. castellanense* Grambast-Fessard, 1969, from the Upper Miocene of France. Of these *A. hamamelidoides*, *H. castellanense* and *H. rhenanum* are distinct because of lower numbers of perforation plate bars, whilst *C. groenlandicus* has a higher number. *H. renaultii* appears to be distinguishable from *H. maegdefrauii* only by its greater vessel-diameter range. This variation is possibly intraspecific, reflecting relative differences in position on the tree (rootwood, branchwood, trunkwood). Until more is known about such variation from a larger sample of the Hungarian material a specific difference based on vessel diameter is probably unwise. *H. maegdefrauii* is, therefore, regarded as *H. renaultii* of which V.63176 is a further record.

Family LAURACEAE Jussieu, 1789

Morphotaxon SABULIA Stopes, 1912

Type species. Sabulia scottii Stopes, 1912, Woburn Sands, Bedfordshire; inferred to have been derived from Upper Pliocene conglomeratic lag phosphorite deposits at the base of the East Anglian Crags, and this in turn derived from uppermost Palaeocene/lowermost Eocene Oldhaven Beds.

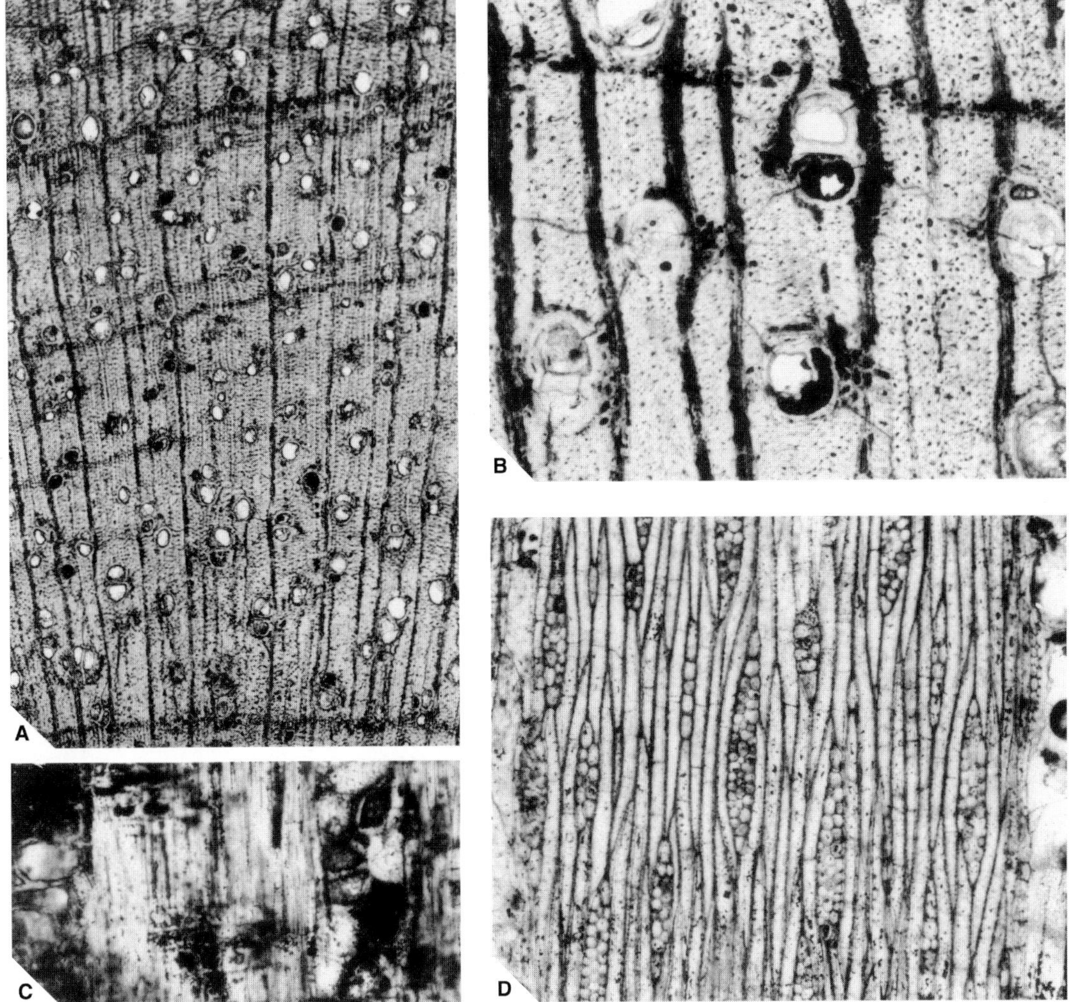

TEXT-FIG. 14. *Sabulia scottii* Stopes, 1912. A, Woburn Sands, Bedfordshire, inferred to have been derived from Upper Pliocene conglomeratic lag phosphorite deposits at the base of the East Anglian Crags, this in turn derived from uppermost Palaeocene/lowermost Eocene Oldhaven Beds. B–D, Upper Pliocene conglomeratic lag phosphorite deposits at the base of the East Anglian Crags, Suffolk; inferred to have been derived from uppermost Palaeocene/ lowermost Eocene Oldhaven Beds. A, TS, showing banded parenchyma, holotype, NHM V.5654c; ×98. B, TS, showing marginal banded and paratracheal parenchyma, paratype, NHM V.63177$1; ×240. C, RLS, scerotic tyloses to left and right of figure, paratype, NHM V.63177$3; ×240. D, TLS, showing rays and septate fibres, paratype, NHM

Sabulia scottii Stopes, 1912

Plate 12; Text-figure 14; Table 9

1912 *Sabulia scottii* Stopes; p. 93, pl. 6, fig. 2; pl. 8, fig. 9.
1915 *Sabulia scottii* Stopes; Stopes, p. 272, text-figs 82–84.
1924 *Sabulia scottii* Stopes; Scott, p. 53.
1931 *Sabulia scottii* Stopes; Edwards, p. 71.
1971 *Sabulia scottii* Stopes; Pant and Kidwai, p. 252.
1989 Dicotyledons Pearson; p. 78.

Holotype. NHM V.5654; slides V.5654a–h.

Paratypes. NHM V.17080; slides V.17080$1–5.NHM V.63177; slides V.63177$1–8.

Other material. NHM, hand specimens: V.64170 (×2), V.64171–72, V.64173 (×4), V.2488, V.2839–40, V.3118, V.4973 (×4), V.4985, V.4991, V.4993, V.4995, V.4999, V.5001–2, V.5004–5, V.5013–14, V.5017, V.5021–23, V.5025, V.5262, V.5279, V.5611, V.5641, V.42049, V.45506, V.53395, V.57420; Slides; V.7816–17, V.8012–14 (V.8013 from hand specimen V.5025), V.8019–30 (V.8024 from hand specimen V.5021), V.8034, V.8039–41, V.8045–47, V.8049–51, V.8053, V.8058 (×3), V.8060, V.8060A, V.8062–67, V.8070. IM, 21 unnumbered hand specimens. SM, hand specimens: C. 84890, C.84891a, K.5251–52, K.5325, one unnumbered specimen presented by Mr J. Carter.

Localities and horizons. Woburn Sands, Bedfordshire; inferred to have been derived from Upper Pliocene conglomeratic lag phosphorite deposits at the base of the East Anglian Crags, and this in turn derived from uppermost Palaeocene/lowermost Eocene Oldhaven Beds (V.5654); Reculvers, Herne Bay, Kent; uppermost Palaeocene/lowermost Eocene Oldhaven Beds (V.17080); Waldringfield, Suffolk (V.42049, V.45506 and all slides except V.8020, V.8030, V.8051, V.8060A, which are only questionably from Suffolk); Woodbridge, Suffolk (V.2839–40, V.8042, V.64171); Felixstowe, Suffolk (V.3118); remainder unlocalised Suffolk (V.5013 Suffolk?) or no locality (V.5262, V.5279, V.5611, V.5641, V.7816–17, V.57420); Upper Pliocene conglomeratic lag phosphorite deposits at the base of the East Anglian Crags; inferred to have been derived from uppermost Palaeocene/lowermost Eocene Oldhaven Beds.

Diagnosis. Vessels solitary and commonly in radial groups of two (50% or more), rarely three. Vessel density is 12–20 mm^2 with an tangential diameters of 59–134 μm. Vessels have simple perforation plates. Pitting is alternate and bordered with horizontal pit apertures of 6–10 μm. Sclerotic tyloses are present in the heartwood. Parenchyma is paratracheal sparse to vasicentric and bluntly aliform, with marginal parenchyma in 1–3 seriate bands and sometimes locally apotracheal banded. Average ray density is 7–8 mm^2, with an average height of 208–260 μm (11–15 cells) and width of 24–35 μm (2–3 cells). Uniseriate rays are very short. Cellular composition is heterogeneous for multiseriate rays, with a single marginal row of upright cells, uniseriate rays composed of square and upright cells, and multiseriate rays with a single marginal row of square to upright cells. The fibres are libriform and 10–25 μm in tangential diameter, often with thick walls.

Description. Three pieces of phosphatized (apatite) secondary wood showing distinct growth ring boundaries (Text-fig. 14A) [**1**] marked by flattened fibres and marginal parenchyma. A central pith is present consisting of large isodiametric cells (Pl. 12, figs 3–4). Primary wood bundles not seen.

Vessels. Vessels diffuse porous (Pl. 12, figs 1–2; Text-fig. 14A) [**5**], solitary 50–70% and in radial groups of 2–3. Perforation plates exclusively simple [**13**]. Vessel to vessel pitting alternate [**22**], bordered, small to medium, 6–10 μm [**25–26**]; vessel to parenchyma pitting possibly with reduced borders, oval [**31**]. Tangential diameter of vessel lumina range 33–176 μm, mean range 59–134 μm [**41–42**]. Density 11–29 mm^2, mean 12–19 mm^2 [**47**]. Vessel element length range 132–330 μm, mean 207 μm [**52**]; sclerotic tyloses present (Pl. 12, figs 5–6; Text-fig. 14C) [**57**].

Imperforate tracheary elements. Septate [**65**] and nonseptate fibres present (Text-fig. 14D) [**66**]; thin to thick-walled [**70**].

EXPLANATION OF PLATE 12

1–8. *Sabulia scottii* Stopes, 1912. 1, 3, 5, 7, Reculvers, Herne Bay, Kent, uppermost Palaeocene/lowermost Eocene Oldhaven Beds. 2, 4, 6, 8, Woburn Sands, Bedfordshire, inferred to have been derived from Upper Pliocene conglomeratic lag phosphorite deposits at the base of the East Anglian Crags, this in turn derived from uppermost Palaeocene/lowermost Eocene Oldhaven Beds. 1, TS, secondary wood, paratype, NHM V.17080$1; ×98. 2, TS, secondary wood, holotype, NHM V.5654b; ×98. 3, TS, primary wood, paratype, NHM V.17080$1; ×98. 4, TS, primary wood, holotype, NHM V.5654c; ×98. 5, TLS, secondary wood, paratype, NHM V.17080$2; ×240. 6, TLS, secondary wood, holotype, NHM V.5654f; ×240. 7, RLS, secondary wood, paratype, NHM V.17080$3; ×98. 8, RLS, secondary wood, holotype, NHM V.5654e; ×98.

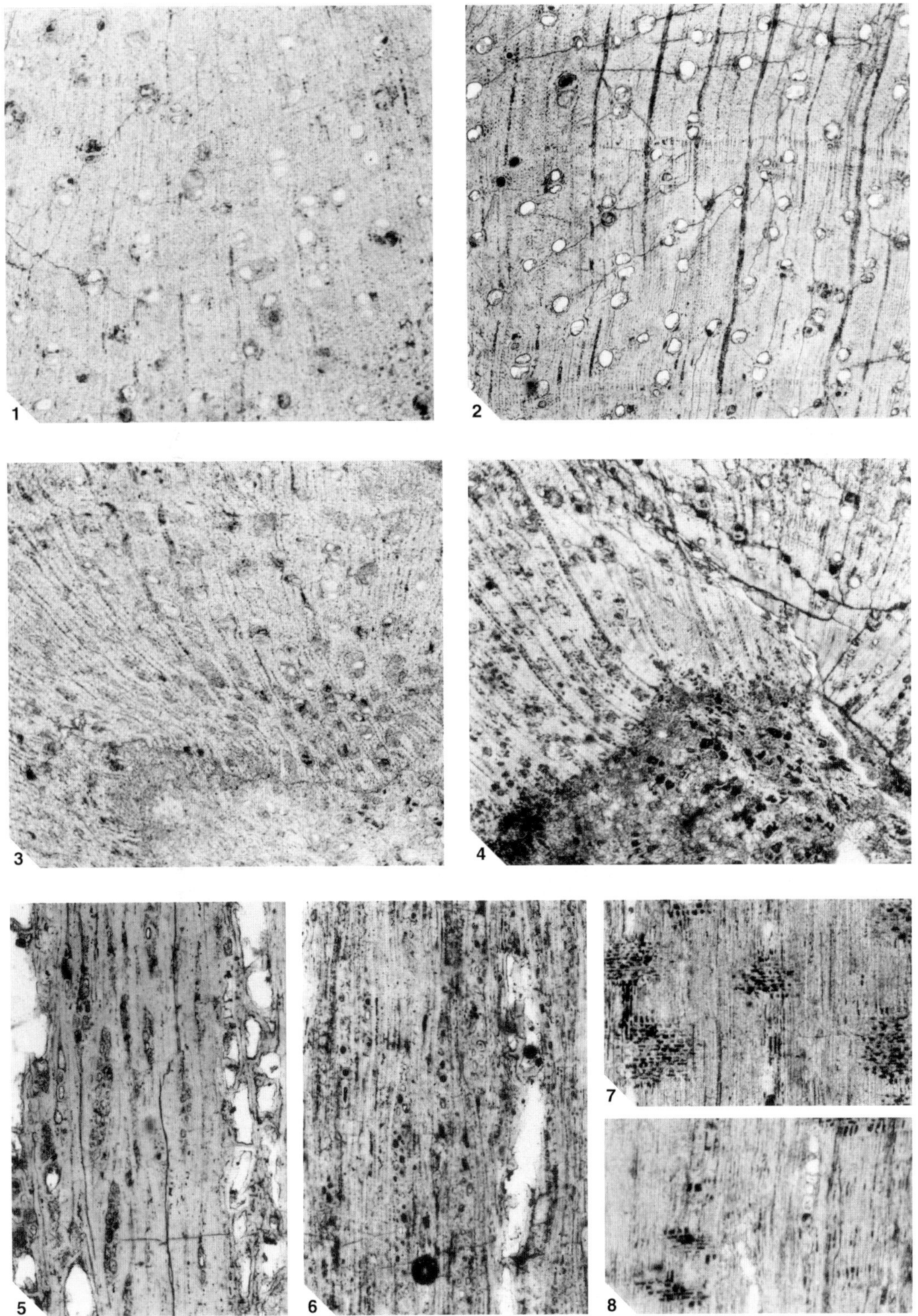

CRAWLEY, *Sabulia*

TABLE 9. *Sabulia scottii*: comparison of the holotype with two other specimens.

	V.5654	V.17080	V.63177
	at least 2·4 cm diam.; primary wood present; no bark; growth rings present, marked by row of flattened fibres and some parenchyma; growth disturbed by emergent branch	at least 6·5 cm diam.; primary wood not preserved; no bark; growth rings present, marked by row of flattened fibres and some parenchyma	at least 6 cm diam.; primary wood not preserved; no bark; growth rings present, marked by row of flattened fibres and some parenchyma
Growth ring boundaries/mm	4·6 3·0 2·5	4·0 3·0 4·0 7·0 6·0	4·5 4·8 5·9 7·8
Vessels	diffuse porous, slightly smaller vessels near growth ring boundary; solitary vessels common, about 50–70%; radial groups of 2 up to 50%, radial groups of 3; sclerotic tyloses present; pitting not seen	diffuse porous, slightly smaller vessels near growth ring boundary; solitary vessels common, about 60%; radial groups of 2 about 25%, radial groups of 3; sclerotic tyloses present; pitting not seen	diffuse porous, slightly smaller vessels near growth ring boundary; solitary vessels common, about 60%; radial groups of 2, 30%, radial groups of 3, 10%; sclerotic tyloses present; intervascular pitting alternate, bordered, 6–10 μm diameter; vessel to parenchyma pitting oval?, up to 10 μm diameter
Vessel density mean per mm^2			
Overall	19·85	19·05	12·15
1st ring	21	20·39	23
2nd ring	20·17	18·17	13
3rd ring	18·75	19·5	9
Tangential diameter μm			
1st ring	58·85	73·15	79·20
2nd ring	64·08	55	106·15
3rd ring	83·88	88	124·85
Rays	mainly 2–3 seriate (exact percentage not determinable); heterocellular with uniseriates very short and composed of square and upright cells; some very tall marginal cells, possibly idioblasts	mainly 2–3 seriate (exact percentage not determinable); heterocellular with uniseriates very short and composed of square and upright cells; some very tall marginal cells, possibly idioblasts	80% rays 2–3 seriate; heterocellular, multiseriates usually with a single marginal row, uniseriates very short and composed of square and upright cells; some very tall marginal cells; possibly idioblasts
Density/mm	8·4	7·1	83
Width μm	32·74	34·62	23·79
Width cells	3	3	2
Multiseriate rays			
Height μm	259·63	266·75	208·45

Height cells	13	15	11
Uniseriate rays			
Height μm	126·5	107·25	118·25
Height cells	6	8	3
Axial parenchyma	paratracheal vasicentric to aliform sheath, appears scanty in places. Possible short bands of local occurrence. Marginal parenchyma possibly present at growth ring boundaries	paratracheal vasicentric to aliform sheath, appears scanty to missing around some vessels. Marginal parenchyma possibly present at growth ring boundaries	paratracheal vasicentric to aliform sheath, appears scanty to missing around some vessels. Marginal parenchyma possibly present at growth ring boundaries
Imperforate tracheary elements	libriform fibres, thin walled	libriform fibres, thin walled (poorly preserved)	libriform fibres, medium to thick walled 10–25 μm (poorly preserved)

Axial parenchyma. Paratracheal, scanty [**78**], vasicentric [**79**], aliform [**80**] winged [**82**], also banded, marginal (all Text-fig. 14A–B) [**89**].

Ray parenchyma. Rays 1–4 cells wide [**97**]; height of multiseriate rays range 66–550 μm (4–30 cells), means 260, 267, 208 μm (13, 15, 11 cells); uniseriate range 55–209 μm (3–7 cells), means 107, 118, 126 (8, 4, 6 cells); ray width range 8–66 μm. Composition is heterogeneous, usually with a single marginal row of upright cells, (Pl. 12, figs 6–7; Text-fig. 14D) [**106**]. Density range 4–12 mm^2, mean 8 mm^2 [**115**].

Oil and mucilage cells. Possible idioblasts at ray margins [**124**].

Comparisons. Stopes did not draw any conclusions about the affinity of *Sabulia*. She stated that 'Every detail of its structure is characteristic of the higher groups of woody Dicotyledons' (1915, p. 276). The vessel distribution, parenchyma and ray structure of these woods is typical of Lauraceae and is closely comparable to the pantropical genus *Cryptocarya*, particularly *C. manii* Hillebrand, 1888, from the Amazon Basin (see Detiénne and Jacquet 1983). Richter (1987), in a comprehensive study of lauraceous wood anatomy, stated that parenchyma distribution is of the highest significance for within-family identification along with vessel pitting, perforations and fibre-type. In all of these, *Sabulia* is closest to *Cryptocarya*. The idioblasts characteristic of this family are not obvious in *Sabulia*, but this is also the case in some species of *Cryptocarya*. About 75 lauraceous-like fossil wood taxa have been described under the names *Cinnamomum*, *Cryptocaryoxylon* Leisman, 1986, *Laurinoxylon* Felix, 1883, *Laurinium* Unger, 1845 and *Ulminium* Unger, 1842. These range in age from Late Cretaceous to Pliocene. Only three of these taxa are comparable with *Sabulia*: *Cryptocaryoxylon gippslandicum* Leisman, 1986, from the Upper Eocene/Lower Oligocene of eastern Victoria, Australia; *Ulminium parenchymatosum* (E. Schönfeld) Crawley, 1989, from the Pliocene of West Germany; and an occurrence of *Laurinoxylon nectandroides* Kräusel and Schönfeld, 1924, recorded by Burgh (1973) from the Miocene Brown Coal of the north Rhineland area of Germany. When compared to *Sabulia*, *Cryptocaryoxylon gippslandicum* has obvious idioblasts amongst the fibres, and *Laurinoxylon nectandroides* has similar vessels, rays and parenchyma but large idioblasts at the ray margins. *Ulminium parenchymatosum* has similar parenchyma but large idioblasts and wider rays, commonly four cells broad.

Remarks. Stopes (1912, 1915) gave a limited description of this wood. Observation is hindered by the presence of sclerotic tyloses in most of the vessels and parenchyma with similar wall thickness to the fibres. The strong transverse walls of Stopes (1912), which she interpreted as possible petrifact, are sclerotic tyloses. There are areas of the holotype that show axial parenchyma distribution as both paratracheal, banded and marginal. It can be detected from the dark cell contents in transverse section and confirmed in longitudinal section. Comparison of the three specimens shows that they are quite uniform in structure (Pl. 12, Table 9) but only V.63177 has patches of really good preservation (Text-fig. 13). This specimen also represents a substantially larger portion of a tree, which accounts for the larger average vessel diameter and vessel element ranges.

Lauraceous remains are recorded throughout the British Palaeogene, essentially from fruit and seed remains (Chandler 1964).

Family TILIACEAE Jussieu, 1789

Morphotaxon TILIOXYLON Burgh, 1978

Type species. *Tilioxylon palaeocordatum* Burgh, 1978, Miocene Brown Coal, northern Germany.

Tilioxylon lueheaformis sp. nov.

Text-figure 15

Derivation of name. For the similarity to Recent genus *Luehea* (Tiliaceae).

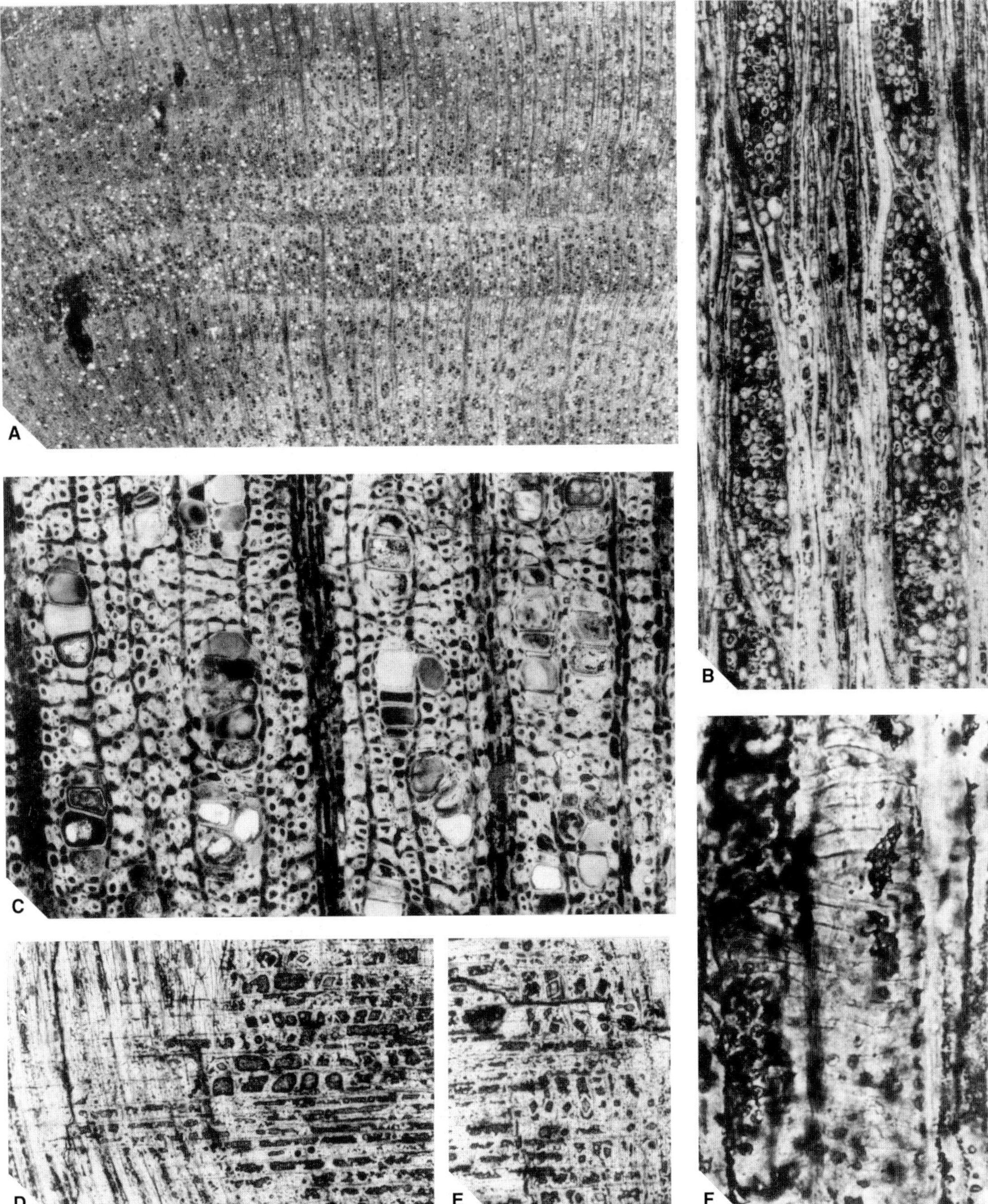

TEXT-FIG. 15. *Tilioxylon lueheaformis* sp. nov., Upper Pliocene conglomeratic lag phosphorite deposits at the base of the East Anglian Crags, Suffolk; inferred to have been derived from uppermost Palaeocene/lowermost Eocene Oldhaven Beds. A, TS, holotype, IM 26$1; ×24. B, TLS, dimorphous rays, paratype, IM 32$3; ×240. C, TS, showing apotracheal diffuse-in-aggregates parenchyma, holotype, IM 26$1; ×240. D, RLS, possible tile cells in upper tiers of ray cells, paratype, IM 32$2; ×300. E, RLS, prismatic crystals in ray cells, paratype, IM 32$2; ×300. F, TLS, helical thickenings in vessel element, paratype, IM 32$3; ×500.

Holotype. IM 26; slides IM 26$1–3.

Paratype IM 32; slides IM 32$1–3.

Locality and horizon. Unlocalised Upper Pliocene conglomeratic lag phosphorite deposits at the base of the East Anglian Crags, Suffolk; inferred to have been derived from uppermost Palaeocene/lowermost Eocene Oldhaven Beds.

Diagnosis. Growth ring boundaries indistinct. Vessels diffuse porous, mean density of 92 mm^2, solitary 4% and in radial multiples of 2–11, mostly 2–6, mean tangential diameter 46 μm, vessel to vessel pitting opposite to alternate, bordered, 5·5–8·5 μm, vessel to parenchyma pitting similar to intervascular. Helical thickenings present in vessel walls. Axial parenchyma apotracheal diffuse-in-aggregates types. Rays of two sizes, uniseriate and 5–6 seriate, multiseriate mean height 1056 μm or 61 cells, uniseriate mean height 349 μm or nine cells, mean ray width 67 μm or 5–6 cells and 12 μm or one cell, heterogeneous, multiseriate rays with 1–4 rows of upright marginal cells, some sheath cells present in wider rays. Prismatic crystals present in ordinary upright cells and chambered procumbent cells. Libriform fibres are present.

Description. Two pieces of an axis with no primary wood. Growth ring boundaries indistinct [**2**], marked by flattened fibres.

Vessel elements. Vessels diffuse porous [**5**] (Text-fig. 15A); solitary vessels 4% and radial multiples of 2–11 vessels [**10**]. Simple perforation plates [**13**]. Intervessel pits opposite [**21**] to alternate [**22**], bordered; pit size small [**25**] to medium [**26**], 5·5–8·5 μm. Vessel to parenchyma pitting similar to intervessel pits [**30**]; pit size 5–9 μm; helical thickenings present [**36**] throughout body of element [**37**] (Text-fig. 15F). Tangential diameter of vessel lumina range 19–77 μm, mean 46 μm [**40**]. Density 72–115 mm^2, mean 92 mm^2 [**49**].

Imperforate tracheary elements. Libriform fibres, no pitting seen [**61**], thin to thick walled [**69**].

Axial parenchyma. Apotracheal, diffuse-in-aggregates [**77**] (Text-fig. 15C).

Ray parenchyma. Ray width 1–8 cells [**98**]; multiseriate ray height range 330–1980 μm (19–116 cells), mean 1036 μm (61 cells) [**102**]; uniseriate ray height range 165–583 μm (4–15 cells), mean 349 μm (nine cells); dimorphous rays present [**103**] (Text-fig. 15B). Multiseriate ray width range 44–94 μm (4–8 cells), mean 67 μm (5–6 cells); uniseriate ray width range 10–14 μm, mean 12 μm. Composition is heterogeneous; body ray cells procumbent with 1–4 rows of upright marginal cells [**107**]; some sheath cells [**110**]; possible *Durio* tile cells present [**111**] (Text-fig. 15D). Density 9–16 rays mm^2, mean 12 mm^2 [**116**].

Mineral inclusions. Solitary prismatic crystals present [**136**] in upright cells [**137**] and in radial alignment in chambered procumbent ray cells [**139**] (Text-fig. 15E).

Comparisons. Important diagnostic characters of this wood are vessels predominantly in radial groups or chains of up to 11 vessels; helical thickening of vessel elements; apotracheal parenchyma of diffuse and diffuse-in-aggregates type; rays of two sizes, uniseriate and five or more; crystals in ray margins and in procumbent cells; tile cells, which are of very limited occurrence in Recent woods (Metcalfe and Chalk 1950). All of these features are seen in Tiliaceae, the fossil described here being most similar to the genus *Luehea*. Similar fossil forms have been referred to the morphotaxa *Grewioxylon* and *Tilioxylon*. All species of *Grewioxylon* possess paratracheal parenchyma and tile cells of the *Pterospermum*-type; these are not present in the Suffolk fossils.

Actinophoroxylon heteroradiatum Kramer, 1974 and *Tilioxylon palaeocordatum* Burgh, 1976 do not have any tile cells. *A. heteroradiatum* has paratracheal and marginal parenchyma, and all rays are multiseriate and storied. *T. palaeocordatum* has similar parenchyma to *A. heteroradiatum* but smaller pitting and homocellular rays. Fossil wood of similar structure to *Luehea* has not been described; hence, the decision here to place the Suffolk fossils in a new species of *Tilioxylon*.

Remarks. Other tiliaceous remains are not known from the Oldhaven Beds but fruits and pollen grains have been recorded from the London Clay Formation (Chandler 1964; Boulter and Hubbard 1982). *Tilioxylon* appears to be another occurrence of a fossil form lacking the storied structure that is usually found in comparable Recent material (Crawley 1989).

Incertae sedis

Morphotaxon DRYOXYLON Schleiden, 1853

Type species. Dryoxylon jenense Schleiden, 1853, Triassic?, Germany.

Dryoxylon calodendrumoides sp. nov.

Text-figure 16

Derivation of name. After the genus *Calodendrum* (Rutaceae), the wood anatomy of which the fossil seems to most resemble.

Holotype. NHM V.63297; slides V.63297$1–3.

Locality and horizon. Unlocalised Upper Pliocene conglomeratic lag phosphorite deposits at the base of the East Anglian Crags, Suffolk; inferred to have been derived from uppermost Palaeocene/lowermost Eocene Oldhaven Beds.

Diagnosis. Growth rings absent. Vessels in oblique to tangential bands, mean density 31 mm^2. Vessels solitary, 30%, radial multiples of 2–5 vessels, mean tangential diameter 74 μm, perforation plates simple. Axial parenchyma paratracheal, scanty. Mean ray density 9 mm^2, width 1–4 cells; mean ray height, multiseriate rays 279 μm or 22 cells, uniseriate rays 121 μm or six cells, heterogeneous with one row of upright marginal cells. Libriform fibres with thick walls.

Description. Primary wood and growth ring boundaries absent [**2**].

Vessel elements. Vessels diffuse porous [**5**], in tangential [**6**] to diagonal pattern [**7**] (Text-fig. 16A–C). Solitary vessels 30% and radial vessel groups of 2–5. Simple perforation plates [**13**]; pitting not seen. Tangential diameter of vessel lumina range 24–116 μm, mean 74 μm [**41**]. Density 25–37 mm^2, mean 31 mm^2 [**48**].

Imperforate tracheary elements. Libriform fibres, no pitting seen [**61**], very thick walled [**70**].

Axial parenchyma. Scanty paratracheal [**78**] (Text-fig. 16D).

Ray parenchyma. Ray width 1–4 cells [**97**]; multiseriate height range 132–748 μm (10–57 cells), mean 279 μm (22 cells); uniseriate height range 44–231 μm (2–12 cells), mean 121 μm (six cells); width 14–55 μm (1–5 cells), mean 35 μm (three cells) (Text-fig. 16E). Composition is heterogeneous; body ray cells procumbent with one row of upright marginal cells (Text-fig. 16B). Density 7–12 rays mm^2, mean 9 mm^2 [**115**].

Comparisons. The oblique to tangential arrangement of the vessels as seen in transverse section is distinctive. Similar vessel distribution can be seen in Recent Fabaceae, Pittosporaceae, Proteaceae and Rutaceae. In Fabaceae the woods have more abundant parenchyma than the fossil described here. Pittosporaceae have similar scanty parenchyma but also possess septate fibres (although these may be few), helical thickening in the vessels, and crystals in marginal ray cells. Proteaceae have much broader rays of up to 30 cells and typically have festoons of parenchyma and vessels between the large rays. In the Rutaceae, *Calodendrum* and *Vepris* both have oblique and tangential vessel patterns, but unlike the fossil there is diffuse and marginal parenchyma, and the rays are homocellular.

The closest fossil species to *D. calodendrumoides* is *Proteoxylon chargeense* Kräusel, 1939, but this is distinguished from it by the tangential rather than radial groups of vessels, the taller rays and axial canals in tangential series.

Remarks. I am uncertain of the affinities of this wood; hence my decision to place it in *Dryoxylon* with the specific epithet to indicate a broad similarity to *Calodendrum* of the Rutaceae.

Discussion of woods from the Oldhaven Beds

The Oldhaven wood assemblage includes five families with two Recent generic analogues that are

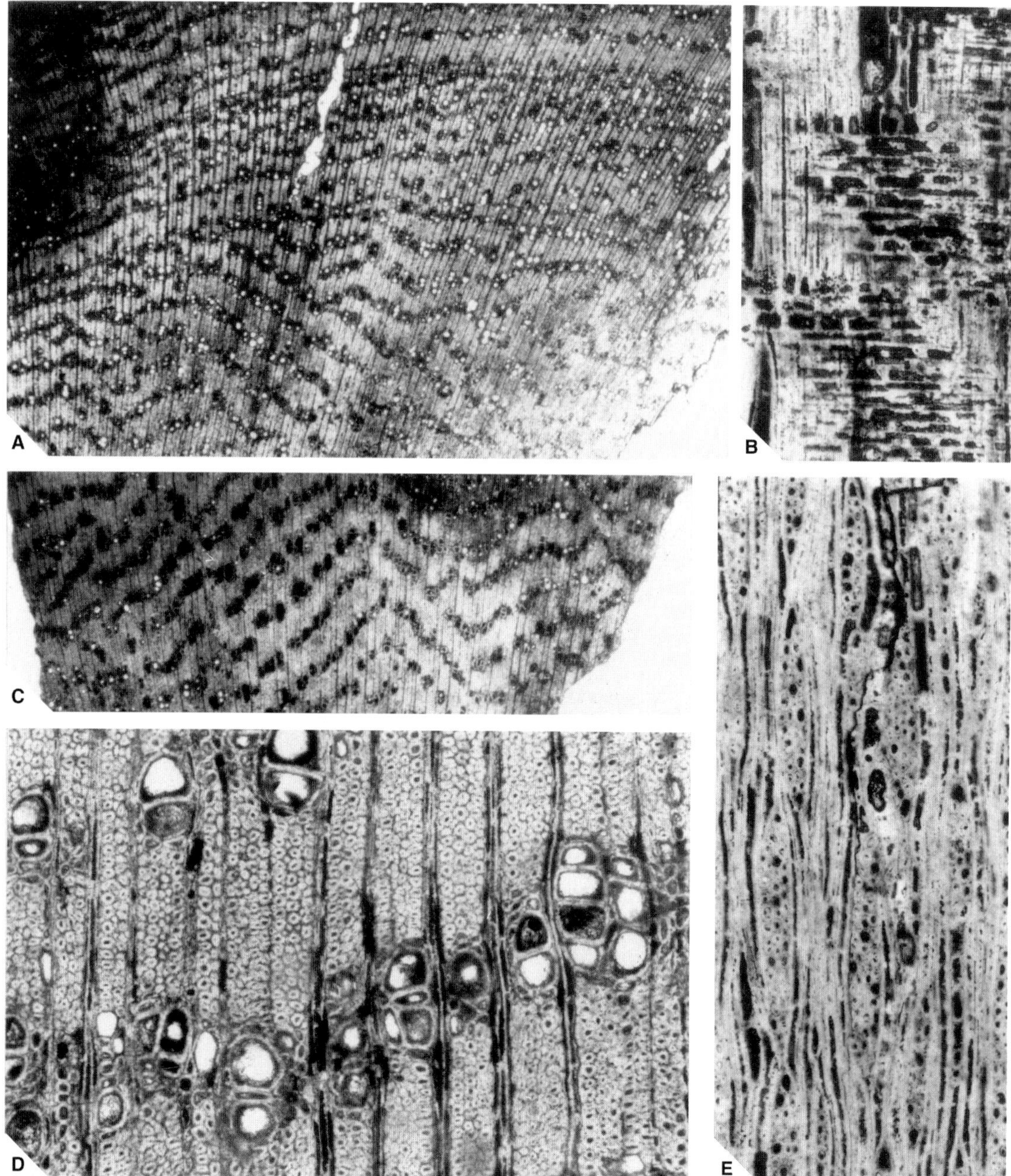

TEXT-FIG. 16. *Dryoxylon calodendrumoides* sp. nov., Upper Pliocene conglomeratic lag phosphorite deposits at the base of the East Anglian Crags, Suffolk; inferred to have been derived from uppermost Palaeocene/lowermost Eocene Oldhaven Beds; holotype, NHM V.63297. A, TS, mainly tangential arrangement of vessels, NHM V.63297$1; ×24. B, RLS, heterogeneous rays, NHM V.63297$3; ×240. C, TS, oblique arrangement of vessels, NHM V.63297$1; ×24. D, TS, notice scanty paratracheal parenchyma, NHM V.63297$1; ×240. E, TLS, multiseriate rays, NHM V.63297$3; ×240.

primarily tropical (see Table 14). Numerically the most important of these is the Lauraceae (*Cryptocarya*-like *Sabulia scottii* Stopes representing at least 50% of all the fossil remains), with Leguminosae (*Bussea*-like), Annonaceae, Flacourtiaceae and Lecythidaceae also present. Another common element, *Plataninium decipiens* (30%), is possibly referable to the Icacinaceae, another pan-tropical family. Other tropical indicators known from the fruit and seed flora include representatives of the Icacinaceae and Menispermaceae (Chandler 1964). Some seasonality of climate is suggested by the diffuse to semi-ring porous structure found in two unidentified wood thin sections (slides V.8020 and V.8033, TS only; Pearson 1989 recorded V.8033 as being ring porous).

Current research on the Oldhaven wood assemblage is restricted to museum collections. Most specimens are from the Crag phosphorite deposits that were extensively quarried in the latter half of the nineteenth century, an activity that ceased in the early part of the last century. Wood remains are very rare in the deposits and are difficult to recognise among more common bone fossils (P. Balson, pers. comm. 1989). Now that the likely source of the fossils is known, collecting should be concentrated on the Oldhaven Beds.

Wood from the Suffolk Pebble Beds

The Suffolk Pebble Beds is thought to be approximately equivalent to the Oldhaven Beds (R. Markham, pers. comm. 1989). It is exposed over a small part of the Ipswich area. The specimen discussed below is the first plant to be described from this formation.

Family STERCULIACEAE (de Candolle) Bartling, 1830

Morphotaxon DOMBEYOXYLON Schenk, 1883

Type species. *Dombeyoxylon aegyptiacum* Schenk, 1883, Oligo-Miocene, Egypt.

Dombeyoxylon cf. *sturanii* Charrier, 1967

Text-figure 17; Table 10

Material. IM R.1990/15; slides IM R.1990/15$1–3.

Horizon and locality. Bramford Brickworks, Suffolk; uppermost Palaeocene/lowermost Eocene Suffolk Pebble Beds; collected by P. R. Crane.

Description. Single piece of silicified secondary wood. Growth ring boundaries indistinct [**2**].

Vessel elements. Vessels diffuse porous [**5**] (Text-fig. 17A); solitary 58% and in radial multiples of 2–3. Simple perforation plates [**13**]. Intervessel pits alternate [**22**], bordered; pit size minute [**24**] (Text-fig. 17D), 2–3 μm. Vessel to parenchyma pits similar to intervessel pits in size and shape [**30**]. Tangential diameter of vessel lumina range 82–220 μm, mean 152 μm [**42**]. Density 6–14 mm^2, mean 10 mm^2 [**47**].

Imperforate tracheary elements. Libriform fibres, no pitting seen [**61**].

Axial parenchyma. Apotracheal, diffuse [**76**] and diffuse-in-aggregates [**77**] (Text-fig. 17C).

Ray parenchyma. Ray width 1–5 cells [**97**]; multiseriate ray height range 363–1320 μm (11–40 cells), mean 589 μm (18 cells); uniseriate ray height range 242–660 μm (5–15 cells), mean 420 μm (nine cells); ray width range 22–132 μm (1–5 cells), mean 71 μm (three cells) (Text-fig. 17B). Composition is heterogeneous; body ray cells procumbent with 1–4 rows of square to upright marginal cells [**107**]. Density 8–13 mm^2, mean 10 mm^2 [**115**].

Comparisons. The notable features of this wood are the entirely apotracheal diffuse-in-aggregates axial parenchyma and minute vessel pitting. It resembles most closely Recent woods of Bombacaceae and Sterculiaceae. In Bombacaceae, however, paratracheal parenchyma is always present, but this is absent from the fossil. In Buettnerioideae (Sterculiaceae), *Mansonia* and *Reevesia* include species that lack

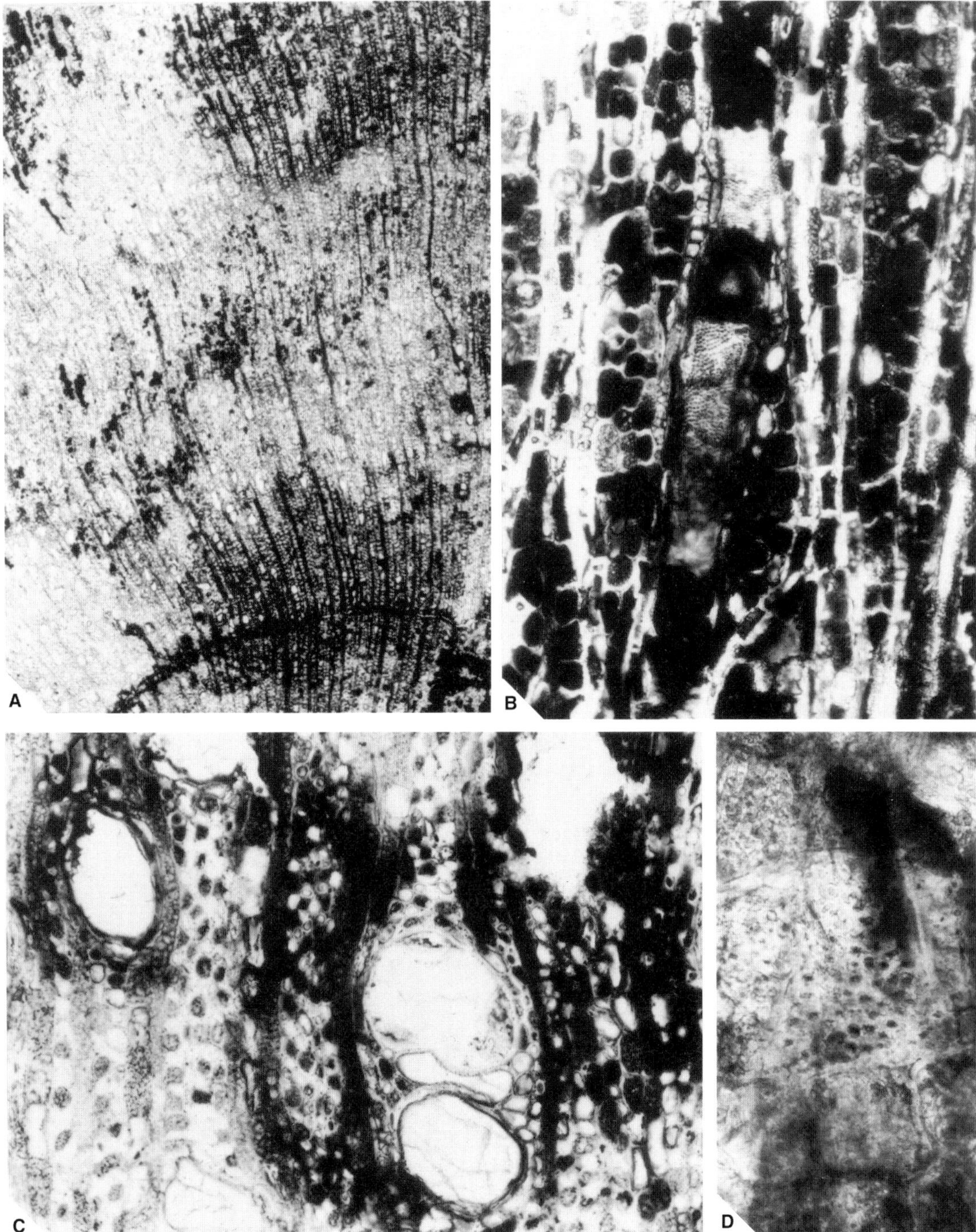

TEXT-FIG. 17. *Dombeyoxylon* cf. *sturanii* Charrier, 1967, Bramford Brickworks, Suffolk, uppermost Palaeocene/lowermost Eocene Suffolk Pebble Beds, IM R.1990/15. A, TS, IM R.1990/15$1; ×24. B, TLS, heterogeneous rays, IM R.1990/15$3; ×350. C, TS, diffuse and diffuse-in-aggregates parenchyma, IM R.1990/15$1; ×280. D, RLS, minute vessel to ray pitting, IM R.1990/15$2; ×800.

TABLE 10. Comparison of *Dombeyoxylon* cf. *sturanii* with *D. sturanii* and *Bombacoxylon owenii*.

	Dombeyoxylon cf. *sturanii* Charrier, 1967 uppermost Palaeocene–lowermost Eocene Pebble Bed	*Dombeyoxylon sturanii* Charrier, 1967 Middle Eocene, Italy	*Bombacoxylon owenii* (Carruthers) Gottwald, 1939 (Privé-Gill and Pelletier 1981) Oligocene, France
Growth rings	indistinct	absent	indistinct
Vessels			
Distribution	diffuse porous solitary	diffuse porous solitary,	diffuse porous solitary,
Grouping	58%, radial groups 2–3	radial groups	radial groups 2–5
Density/mm^2			
range	6–14	–	3–5
mean	10	6	–
Tangential diameter μm			
range	82–220	50–150	37–185
mean	152	–	–
Element length μm			
range	–	350–500	160–330
mean	–	–	–
Perforation plate	simple	simple	simple
Pitting			
intervascular	alternate, bordered 2–3 μm	alternate, bordered 3 μm	alternate, bordered 6–9 μm
vessel to parenchyma	as above	as above	as above
Axial parenchyma	apotracheal, diffuse and diffuse-in-aggregates	apotracheal, diffuse and diffuse-in-aggregates	apotracheal, diffuse and diffuse-in-aggregates
Ray parenchyma			
Density/mm			
range	8–13	–	10–13
mean	10	8	–
Width μm			
range	22–132	–	–
mean	71	–	–
Multiseriate ray width cells			
range	1–5	1–3	1–3
mean	2–3	–	–
Multiseriate height μm			
range	363–1320	300–370	50–410
mean	579	–	–
Multiseriate height cells			
range	11–40	–	2–19
mean	18	–	–
Uniseriate height μm			
range	242–660	–	–
mean	420	–	–
Uniseriate height cells			
range	5–15	–	–
mean	9	–	–
Composition	heterocellular, 1–4 marginal rows of square to upright cells	homocellular	weakly heterocellular
Imperforate tracheary elements	libriform fibres	libriform fibres	libriform fibres
Mineral inclusions	–	crystals in chambered axial parenchyma	–

paratracheal parenchyma but the small rays are always storied, again a feature lacking from the fossil. Comparable fossils are *Bombacoxylon owenii* (Carruthers) Gottwald, 1969 (Bombacaceae), and *Dombeyoxylon sturanii* Charrier, 1967 (Sterculiaceae) (see Table 10). *B. owenii* has larger (6–9 μm) pitting. *Dombeyoxylon sturanii* has smaller vessels, homocellular rays and crystalliferous parenchyma.

Remarks. Other sterculiaceous fossil remains are known from the Reading Formation (wood, Crawley 1989), London Clay Formation (*Sphinxia ovalis* Reid and Chandler, 1933 and pollen grains: Chandler 1964; Boulter and Hubbard 1982). Bombacaceous remains are so far unknown from the British Palaeogene. The family has a tropical to subtropical distribution today. Storied structure is typical of Recent Sterculiaceae but is not present in the fossil. Crawley (1989) and Manchester (1979, 1980) have both shown that other Eocene sterculiaceous woods lack this feature, whereas comparable Recent analgoues have it.

Discussion of wood from the Suffolk Pebble Beds

Plant remains appear to be very rare in the Suffolk Pebble Beds. Even small fragments of wood are uncommon in residues (Dr J. J. Hooker, pers. comm. 1996). *Dombeyoxylon* cf. *sturanii* is the only plant to have been described from them. The wood structure is typical of tropical sterculiaceous trees, and as such is consistent with the subtropical to tropical conditions that prevailed in southern England during the Early Eocene.

The fossil mammal fauna from this bed is very similar to that of the Reading Formation and is possibly derived from it (Dr J. J. Hooker, pers. comm. 1996). The preservation of *Dombeyoxylon* cf. *sturanii* is unlike that of Reading Formation woods although both are silicified. I have no reason to suspect that this specimen is not *in situ*.

Woods from the London Clay Formation

The London Clay Formation has long been famous for abundant macrofossil remains of plants, particularly on the Isle of Sheppey, from which they have been collected since 1705 (Reid and Chandler 1933). The diversity of this flora was revealed by Bowerbank (1840), Reid and Chandler (1933) and Chandler (1961, 1964, 1978). To date 143 taxa from 19 families have been described, of which the majority are angiosperm fruits and seeds (Collinson 1983). This diverse flora, which seems to include tropical, subtropical and temperate plants, has generated a great deal of interest in the climate and environment during the deposition of the London Clay. Many of the fruits and seeds regarded as tropical were thought to be comparable to those produced by the vegetation of present-day Indo-Malaysia (Reid and Chandler 1933) and the south-eastern USA (Chandler 1961, 1964). The temperate elements, including gymnosperms, were assumed to have grown on nearby higher ground with the tropical elements in the lowlands. However Daley (1972) disputed the existence of any high land surrounding the depositional basin and proposed a frostless climate with the tropical elements living in shallow valleys with a high mean humidity. Collinson (1983) regarded the closest Recent analogue as paratropical rainforest (Wolfe 1979) owing to the following features of the flora: (1) the presence of elements in the macroflora and microflora that are often thought of as temperate; (2) the precise combination of plants in the macroflora; (3) the seasonality indicated by growth rings; (4) the absence of important rain forest elements like dipterocarps; and (5) the proposed oxygen isotope temperatures of between 20–27°C for the Lower Eocene. These conclusions were based primarily upon studies of the fruit, seed and pollen remains. Fossil wood remains were under-used in this assessment although they form the largest constituent of the macroflora. They occur as calcified or lignified 'logs' and pyritized 'twigs' ('twigs' here include angiosperm and conifer twigs, tubers, rhizomes, and fern rachides). The logwood, although often consisting of a high proportion of 'ship-worm' burrows (*Teredo* spp.), can be well preserved (Seward 1919; Brett 1956, 1960, 1972; Crawley 1989) and family affinities (Anacardiaceae, Cupressaceae, Cercidiphyllaceae, Fagaceae and Sapotaceae) recognised. A preliminary study of the 'twig' flora by Scott and de Klerk (1974) showed that this material can yield useful data. More recently, work has been done on the anatomy of embryonic mangrove hypocotyls and dicotyledonous wood twigs (Rhizophoraceae, Celastraceae, Dipterocarpaceae, Icacinaceae? or Platanaceae? and Sapindaceae: Wilkinson 1981, 1983, 1984, 1988; Poole 1992, 1993; Poole and Wilkinson 1992, 1999) and on fern rachides (Collinson and Ribbins 1977; Ribbins and Collinson 1978).

Family MELIACECAE Jussieu, 1789

Morphotaxon MELIACEOXYLON Greguss, 1969

Type species. Meliaceoxylon matrense Greguss, 1969, Miocene, Hungary.

Meliaceoxylon collinsonae sp. nov.

Plate 13

Derivation of name. After the collector Dr M. E. Collinson.

Holotype. NHM V.63463; slides V.63463$1–7.

Locality and horizon. Foreshore, Warden Point, Isle of Sheppey, Kent; Lower Eocene London Clay Formation, Unit B; collected by Dr M. E. Collinson in 1989.

Diagnosis. Growth rings absent. Vessels diffuse porous, mean density 10 mm^2, solitary 74% and in radial multiples of 2–4 vessels, mean tangential diameter 121 μm, perforation plates simple, all pitting alternate, bordered and 4–5 μm. Axial parenchyma paratracheal, vasicentric to lozenge aliform and locally banded, up to four cells broad. Ray mean density 10 mm^2, mean width 44 μm, range 1–5 cells, mean ray height multiseriate 518 μm or 24 cells, uniseriate 167 μm or five cells, heterogeneous with 1–2 rows of upright marginal cells. Solitary rhomboidal crystals in both upright and procumbent ray cells.

Description. Primary wood and growth ring boundaries are absent (Pl. 13, fig. 1) [**2**].

Vessel elements. Vessels diffuse porous (Pl. 13, fig. 1) [**5**], solitary 74% and in radial groups of 2–4 vessels. Simple perforation plates [**13**]. Intervessel pits alternate [**22**], bordered, small; pit size 4–5 μm. Vessel to parenchyma pits similar to intervessel pits; pit size 4 μm. Tangential diameter of vessel lumina range 77–148 μm, mean 121 μm [**42**]. Density 8–14 mm^2, mean 10 mm^2 [**47**].

Imperforate tracheary elements. Septate fibres present [**65**], no pitting seen, very thin walled [**68**].

Axial parenchyma. Paratracheal, vasicentric [**79**] to aliform (Pl. 13, fig. 3)[**80**], confluent [**83**]; also banded; bands up to nine cells wide (Pl. 13, fig. 4) [**85**].

Ray parenchyma. Ray width 1–5 cells [**97**]; multiseriate ray height range 143–957 μm (6–42 cells), mean 518 μm (24 cells); uniseriate ray height range 66–275 μm (2–8 cells), mean 167 μm (five cells); ray width 11–73 μm (1–5 cells), mean 44 μm (three cells). Composition is heterogeneous (Pl. 13, fig. 2); body ray cells procumbent with usually one (Pl. 13, fig. 5) [**106**] but sometimes up to four rows of upright marginal cells [**107**]. Density 7–12 rays mm^2, mean 10 mm^2 [**115**].

Mineral inclusions. Solitary prismatic crystals present (Pl. 13, fig. 5) [**136**] in upright cells [**137**] and procumbent ray cells [**138**].

EXPLANATION OF PLATE 13

1–5. *Meliaceoxylon collinsonae* sp. nov., foreshore, Warden Point, Isle of Sheppey, Kent, Lower Eocene London Clay Formation, Unit B, holotype, NHM V.63463. 1, TS, *Teredo* boring to left and at bottom of figure, NHM V.63463$1; ×24. 2, TLS, heterogeneous rays, NHM V.63463$3; ×240. 3, TS, paratracheal vasicentric to lozenge aliform, NHM V.63463$1; ×240. 4, TS, banded parenchyma, NHM V.63463$1; ×240. 5, RLS, prismatic crystals in ray cells, NHM V.63463$2; ×800.

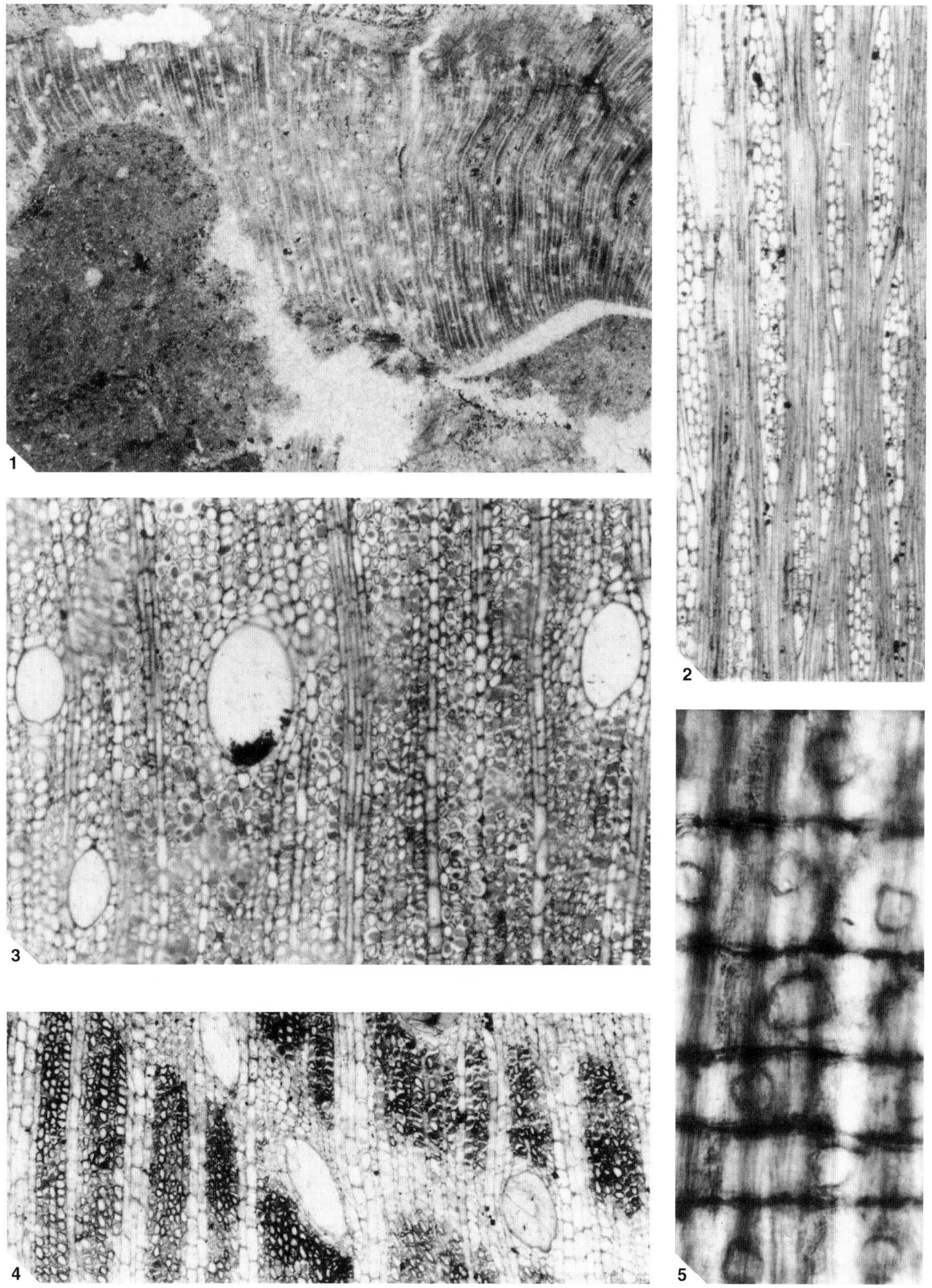

CRAWLEY, *Meliaceoxylon*

Comparisons. The features of this fossil are most similar to Recent woods of the Combretaceae and Meliaceae. In the Combretaceae the genus *Terminalia* is most similar to the fossil but small pitting is not typical and only occurs in a single species (van Vliet 1979). Crystals also occur in enlarged cells or idioblasts; these are not present in the fossil. In the Meliaceae, all features of the fossil are present but not in a single Recent genus. The features span the anatomy of *Entandrophragma* and *Khaya*, with axial parenchyma of the type found in the former and rays similar to those of the latter. Fossil forms close to *Meliaceoxylon collinsonae* sp. nov. are: *Carapoxylon ornatum* (Felix) Mädel, 1960, from the Upper Miocene of Germany; *C. cahenii* Lakhanpal and Prakash, 1970, from the Miocene of Africa; *Entandrophragminium magnieri* (Louvet) Prakash, 1976, from the Oligocene of Libya; *E. lateparenchimatosum* (Petrescu) comb. nov., from the Oligocene of Romania; and *E. normandii* (Louvet) comb. nov., from the Upper Eocene/Lower Oligocene of Algeria. The Sheppey wood is distinct from *Entandrophragminium* species because the rays of the former are heterogeneous with 1–4 marginal rows of square to upright cells; and from *Carapoxylon* because the axial paratracheal parenchyma is vasicentric to lozenge aliform and not scanty paratracheal. This new species is, therefore, referred to *Meliaceoxylon* Greguss, 1969, and is intended for woods of probable meliaceous affinity but which are not closely comparable to any Recent genus.

Remarks. Two genera and species of fruit of Meliaceae have been recorded from the London Clay Formation (Chandler 1964). The two Recent genera that are most similar to the fossil, namely *Entandrophragma* and *Khaya*, grow in the forests of tropical Africa (Kribs 1959).

Family PLATANACEAE? Dumortier, 1829 or ICACINACEAE? (Bentham) Miers, 1851

Morphotaxon PLATANINIUM Unger, 1842, emend. Page, 1968

Type species. Plataninium acerinum Unger, 1842, horizon and locality unknown.

Plataninium decipiens Brett, 1972

Table 11

1972 *Plataninium decipiens* Brett, p. 497, pl. 99, figs 3–6.
1977 *Plataninium decipiens* Brett, Süss and Müller-Stoll, pp. 50–51, 57.
1989 *Plataninium decipiens* Brett, Crawley, p. 612.
1989 *Plataninium decipiens* Brett, Pearson, p. 76.

Holotype. NHM V.45684; slides V.45684a–d.

Paratype. NHM V.45685; slides V.45685a–c.

Other material. NHM as Pearson 1989 P. 77; IM: hand specimens 19, 20; SM: hand specimens; CMU, three unnumbered specimens, one labelled proteaceous wood, Woodbridge, Wiltshire Collection.

Locality and horizon. Isle of Sheppey, Lower Eocene London Clay Formation, Unit B (V.45684); Herne Bay, Kent; uppermost Palaeocene/lowermost Eocene Thanet Sand Formation (V.45685).

Remarks. Brett (1972) mentioned that this species might prove synonymous with *Plataninium europeanum* Prakash, Brezinova and Bužek, 1971, with which it showed great similarity (Pearson 1989 reiterated this). In Table 11 the holotype and paratype of *Plataninium decipiens* are compared with *P. europeanum*. In quantitative features the specimens are very close. They differ because *P. europeanum* has libriform fibres, not fibre-tracheids, and has alternate to transitional pitting, not opposite to scalariform as in *P. decipiens*. The variation in pitting may be intraspecific (see *Cantia arborescens*); however, the difference in ground tissue elements cannot be similarly explained. The

TABLE 11. Comparison of *Plataninium decipiens* with *P. europeanum*.

	Plataninium europeanum Prakash, Brezinova and Buzek, 1971	*Plataninium decipiens* Brett, 1972, holotype, V.45684 Lower Eocene London Clay Formation, Isle of Sheppey, Kent	*Platanium decipiens* Brett, 1972, paratype, V.45685 Upper Palaeocene Thanet Sand Formation, Herne Bay, Kent
Growth rings	distinct	indistinct	indistinct
Vessels			
Distribution	diffuse porous solitary	diffuse porous solitary	diffuse porous solitary,
Grouping	irregular groups 2–3		rare tangential groups 2–4
Density/mm^2			
range	32–65	–	–
mean	–	30	42
Tangential diameter μm			
range	28–128	35–105	54–120
mean	–	80	88
Element length μm			
range	–	–	–
mean	–	–	–
Perforation plate	scalariform, 13–30 bars	scalariform, 13–25 bars,	scalariform, 17–32 bars, mean 25
Pitting			
intervascular	alternate to opposite and transitional; 6–12 μm diameter	opposite to scalariform	opposite to scalariform
vessel to parenchyma	–	–	–
Axial parenchyma	apotracheal, diffuse and diffuse-in-aggregates	apotracheal, diffuse and diffuse-in-aggregates	apotracheal, diffuse and diffuse-in-aggregates
Ray parenchyma			
Density/mm			
range	1–13	1–4	1–3
mean	2	3	–
Multiseriate ray width μm			
range	64–360	–	–
mean	–	–	–
Multiseriate ray width cells			
range	3–16	2–18	4–25
mean	10	6	–
Multiseriate height μm			
range	1000+	1000–6000	up to 5000
mean	–	–	–
Multiseriate height cells			
range	–	–	–
mean	–	–	–
Uniseriate height μm			
range	–	–	–
mean	–	–	–
Composition	heterocellular to heterocellular with 1–2 marginal rows of procumbent or square to upright cells	homocellular, 1–2 marginal rows	homocellular, 1–2 marginal rows of square cells
Imperforate tracheary elements	libriform fibres	fibre-tracheids	fibre-tracheids
Mineral inclusions	–	solitary crystals in ray parenchyma	–

apparent difference in fibre pitting might reflect differences in preservation, but until this can be clarified the species should remain separate.

Family RUTACEAE Jussieu, 1789

Morphotaxon FAGAROXYLON Burgh, 1964

Type species. Fagaroxylon limburgense Burgh, 1964, Miocene, Limburg, The Netherlands.

Fagaroxylon atkinsoniae (Crawley) comb. nov.

Table 12

1989 *Sapotoxylon atkinsoniae* Crawley; p. 614, pl. 69, figs 7–9; pl. 70, figs 1, 9; table 5.
1991 *Sapotoxylon atkinsoniae* Crawley; Selmeier; pp. 68–69.

Locality and horizon. Foreshore, Isle of Sheppey, Kent; Lower Eocene London Clay Formation, Unit B.

Emended diagnosis. As stated originally except that the axial parenchyma comprises marginal bands 1–4 cells wide.

Remarks. Comparative work on this taxon has shown that the banded parenchyma of this wood type to be marginal parenchyma and that there are also scarce paratracheal parenchyma cells abutting vessels. These features are characteristic of the Rutaceae. These fossils have radial vessel multiples that retain a rounded shape, another feature of rutaceous wood. Genera such as *Balfourodendron, Chloroxylon, Esenbeckia* and *Teclea* resemble *Fagaroxylon atkinsoniae* comb. nov. in most qualitative features. These Recent wood analogues are found in tropical regions of the Southern Hemisphere.

Fagaroxylon (Rutaceae) contains two species similar to the Sheppey material (Table 12). *Fagaroxylon bavaricum* Selmeier, 1975 has a higher percentage of solitary vessels (82%) and *F. limburgense* Burgh, 1964 has marginal bands one cell thick and denser, higher rays.

Fourteen species of rutaceous fruits and seeds and one pollen type have been described from the London Clay Formation (Chandler 1964).

Discussion of woods from the London Clay Formation

The woods preserved in the London Clay Formation are further examples of the dominant tropical element of the vegetation at the time of deposition, an estimated 92% of the flora (Collinson 1983). They are also consistent with other records of the Meliaceae and Rutaceae.

Wood from an unknown horizon, Woburn Sands, Bedfordshire

Family DIPTEROCARPACEAE Blume, 1825

Morphotaxon DIPTEROCARPOXYLON Holden, 1916, emend. Den Berger, 1927

Type species. Dipterocarpoxylon schenkii (Felix) Schweitzer, 1958, Tertiary, Java.

Dipterocarpoxylon porosum (Stopes) Kräusel, 1922

Text-figure 18

1912 *Woburnia porosa* Stopes, p. 92, pl. 7, fig. 7; pl. 8, fig. 8.

TABLE 12. Comparison of *Fagaroxylon atkinsoniae* with other *Fagaroxylon* species.

	Fagaroxylon atkinsoniae (Crawley) comb. nov. Lower Eocene, Kent V.62704–6	*Fagaroxylon bavaricum* Selmeier (1975) Upper Miocene, Germany	*Fagaroxylon limburgense* van der Burgh (1964, 1973) Miocene, Germany
Growth rings	Indistinct	Indistinct	Indistinct
Vessels			
Distribution	Diffuse porous	Diffuse porous	Diffuse porous
Grouping	Solitary 30%, radial groups 2–4	Solitary 82%, radial groups 2–6	Solitary, radial groups
Density/mm^2			
range	20–25	32–55	15
mean	–	44	–
Tangential diameter μm			
range	42–140	28–114	60–100
mean	63/80/98	70	–
Element length μm			
range	250–560	93–372	–
mean	300/400/420	203	–
Perforation plate	Simple	Simple	Simple
Pitting			
intervascular	Alternate bordered 3–5 μm	Sub-alternate to alternate, bordered 3–5 μm	Alternate
vessel to parenchyma	As above	–	As above
Axial parenchyma	Paratracheal, scanty apotracheal (rare isolated cells), diffuse and marginal (1–3 cells wide)	Paratracheal, scanty apotracheal, diffuse and marginal	Paratracheal, scanty apotracheal marginal
Ray parenchyma			
Density/mm			
range	7–9	5–12	10–16
mean	–	7	–
Multiseriate ray width μm			
range	15–30	18–71	–
mean	22/23/25	–	–
Multiseriate ray width cells			
range	1–2	1–6	1–4
mean	–	5	–
Multiseriate height μm			
range	180–980	168–1130	–
mean	280/560/580	–	–
Multiseriate height cells			
range	6–35	13–72	–
mean	10/12/23	–	20/35
Uniseriate height μm			
range	110–600	28–260	–
mean	150/200/–	–	–
Uniseriate height cells			
range	2–14	1–17	–
mean	4/7/–	–	–
Composition	heterocellular, 1–4 rows of square to upright marginal cells	homocellular	heterocellular, single rows of square marginal cells
Imperforate tracheary	libriform fibres	libriform fibres	libriform fibres
Mineral inclusions	solitary crystals in chambered parenchyma cells	solitary crystals in chambered parenchyma cells	solitary crystals in chambered parenchyma cells

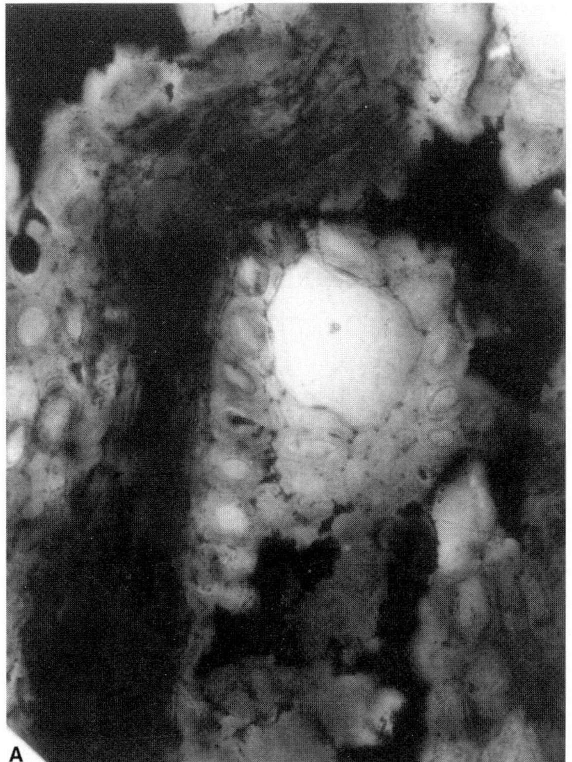

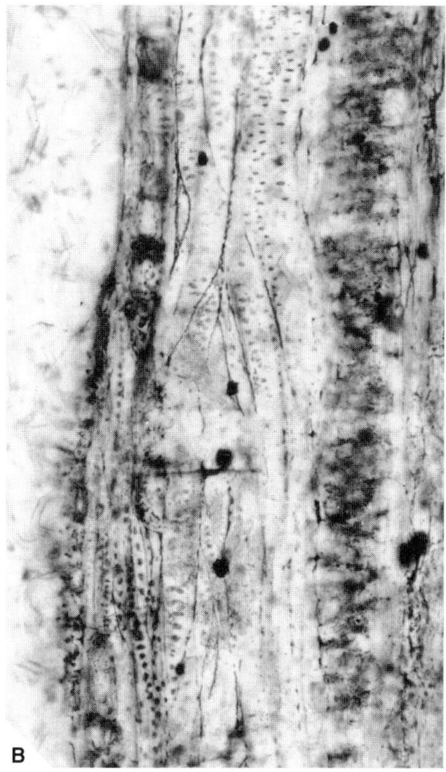

TEXT-FIG. 18. *Dipterocarpoxylon porosum* (Stopes) Kräusel, 1922, Woburn Sands, Bedfordshire, age unknown, holotype, NHM V.5452. A, TS, vertical intercellular canal; the epithelial lining can be clearly seen, NHM V.5452h; ×1200. B, TLS, vasicentric tracheids, NHM V.5452h; ×480.

1915 *Woburnia porosa* Stopes; Stopes, pp. 26–28, text-figs 79–81.
1922 *Dipterocarpoxylon porosum* (Stopes) Kräusel, pp. 12, 14.
1933 *Woburnia porosa* Stopes; Bancroft, pp. 86–87.
1958 *Dipterocarpoxylon porosum* (Stopes) Kräusel; Schweitzer, pp. 19–20, pl. 2, figs 5–6; pl. 3, figs 1–2.
1971 *Woburnia porosa* Stopes; Pant and Kidwai, p. 252, text-fig. 3c.
1981 *Woburnia porosa* Stopes; Page, p. 445.

Holotype. NHM V.5254; slides V.5254a–i.

Locality and horizon. Woburn Sands, Bedfordshire; age unknown.

Remarks. As no further specimen has come to light and there is no adherent matrix, only a short discussion of its affinities and possible origin is presented here. The specimen is from Woburn Sands, Bedfordshire (Stopes 1912). It was subsequently placed in *Dipterocarpoxylon* (Dipterocarpaceae) by Kräusel (1922) and Schweitzer (1958). Neither Bancroft (1933) nor Page (1981) were convinced that it had axial canals characteristic of this family but these canals are definitely present if not abundant (Text-fig. 18A). A search of the Oxford\PRNL\NCSU database using the features of *Dipterocarpoxylon porosum* [vessels exclusively solitary; simple perforations; parenchyma sparse; rays up to 4–8 cells wide with a few uniseriates and often more than 1 mm high; vasicentric tracheids present (Text-fig. 18B); and ordinary axial canals present] led to *Dipterocarpus* spp. being the closest match. This supports the views of Kräusel and Schweitzer.

Many of the vessels in the holotype show evidence of chalcedony (D. T. Moore, pers. comm. 1989). I have only seen this type of preservation in one other specimen (*Fagoxylon* sp., V.63172; see Woods from the Reading Formation). The specimens are also similar in that both are silicified and show no evidence of borings. Woods from the Reading Formation are characteristically cream to pale grey, sometimes with a darker core, often with much evidence of iron impurities. The hand specimen of *Dipterocarpoxylon porosum* is darker grey than any Reading wood I have seen, but of the silicified wood horizons known in the British Cretaceous and Tertiary succession, the sandy facies of the Reading Formation would currently seem the most likely source of the specimen.

In her report of 1970 to the Palaeobotany Section of The Natural History Museum, Prof. V. M. Page mentioned seeing a *Woburnia*-like wood from the Glyme Valley, near Woodstock, Oxfordshire. It was said to have been picked up as a float and the youngest deposits in that area are Jurassic (V. M. Page, pers. comm. 1989).

DISCUSSION AND CONCLUSIONS

Stopes' putative 'Lower Cretaceous angiosperm woods'

Aptiana radiata Stopes 1912 is accepted as an angiosperm wood following thorough comparison with wood of both Gnetales and dicotyledons. It is considered to be from the Aptian/Albian (Lower Greensand), probably from the Folkestone Beds of Luccomb Chine, Isle of Wight or Copt Point, Kent. There are no new specimens and it is considered to be the only known British Cretaceous angiosperm wood.

Examination of new material of the following three taxa found in the collections of The Natural History Museum has led to the following observations: (1) *Cantia arborescens* Stopes, 1915 is assigned to the Betulaceae and the holotype is derived from the Thanet Sand Formation (Palaeocene); (2) *Hythia elgarii* Stopes, 1915 is considered to be referable to the Fagaceae?, Icacinaceae? or Platanaceae? and the holotype is derived from the Thanet Sand Formation; (3) *Sabulia scottii* Stopes, 1912 is assigned to Lauraceae, and the holotype is derived from the Oldhaven Beds (uppermost Palaeocene/lowermost Eocene).

No new specimens of *Woburnia porosa* Stopes, 1912 have been found. It is accepted as a species of *Dipterocarpoxylon* (Dipterocarpaceae). Its age is still problematic but it is likely to be Tertiary.

British Palaeogene trees and vegetation

All the wood remains in this study were subjected to fluvial or marine dispersal. Those of the fluvial Reading Formation are probably more indicative of local conditions than those of the marine Palaeogene sediments. The woods of the Oldhaven Beds are unlike those from the Thanet Sand and London Clay formations, which are often heavily bored by marine molluscs (e.g. *Teredo* spp.), evidence of substantial periods of immersion and probable widespread dispersal. The majority of the woods described have no Recent equivalent at generic level and probably represent extinct taxa. The plants concerned may have flourished under conditions for which there is no modern equivalent (Creber and Chaloner 1984) and/or the correlation of wood anatomy with ecology may not have been constant over time (Wheeler and Baas 1991).

A comparison of wood with other types of plant remains from the British Palaeogene is shown in Table 13. The Fagaceae have a wide distribution, with occurrences in the Interbasaltic Beds of Rhum, Thanet Sand Formation, the Oldhaven Beds? and the London Clay Formation (Tomkeieff and Blackburn 1942; Brett 1960; Chandler 1964). The wood, leaf and pollen remains are of *Castanea*, *Castanopsis*, *Lithocarpus* (all evergreen oaks), *Fagus*, *Nothofagus* or *Quercus*-like plants. Icacinaceae/Platanaceae are similarly represented. More families occur in the London Clay Formation than in the other formations discussed. Much of the floral continuity with older deposits is apparent in wood remains, highlighting their usefulness.

According to Collinson and Hooker (1987) the Palaeogene succession in southern England contains floristically distinct fossil assemblages. Those of the Upper Palaeocene are less diverse and have a reduced tropical aspect, whereas Early Eocene–early Middle Eocene floras are more diverse with an increased tropical component. The greater diversity in the Oldhaven woods is noteworthy compared to that of the

TABLE 13. Family distribution of fossil remains across selected formations in the British Palaeogene: F, fructification; L, leaf; P, pollen; W, wood (including 'twig' wood); information from Brett (1956, 1960, 1966, 1972), Chandler (1964), Ward (1978), Collinson (1983), Collinson and Hooker (1987), Boulter and Kvaček (1989), Crawley (1989), Poole (1992, 1993), and Poole and Wilkinson (1999).

Family	Brito-Arctic Igneous Province	Thanet Sand Formation	Reading Formation	Oldhaven Beds	London Clay Formation
Anacardiaceae	–	W	W?, F?	–	W, P
Annonaceae	–	–	–	W	F
Apocynaceae	–	–	W?, F?	–	F
Aquifoliaceae	–	W	–	–	F, P
Betulaceae	–	W	F, L P	–	P
Buseraceae	–	–	W?	–	F, P
Caesalpiniaceae	–	–	F, L?	W	W, F
Celastraceae	–	–	–	–	W
Cercidiphyllaceae	W, L	–	F, L	W	W, L
Cornaceae	–	–	F	W?, F?	F, P
Cunoniaceae	W?	–	–	W?	–
Dipterocarpaceae	–	–	–	–	W
Eucryphiaceae	W?	–	–	–	–
Euphorbiaceae	–	W	–	F?	F
Eupteleaceae	–	–	–	W?	–
Fagaceae	L?, P?	W	W	W?	W, P
Flacourtiaceae	–	–	F?	W	F
Hamamelidaceae	–	–	F?	W	F
Icacinaceae	W?	W, F	W?, F	W?	F, P
Lauraceae	–	W?	W?, F, L	W, F	F
Lecythidaceae	–	–	–	W	–
Meliaceae	–	–	W	–	W, F
Platanaceae	W?, L, P	W?	W?, L	W?	W?
Rutaceae	–	–	F	W?, F	W, P
Sapotaceae	–	–	–	W	F, P
Sapindaceae	–	–	–	–	W
Sterculiaceae	–	–	W	–	F, P
Theaceae	–	–	F	W?, F	F
Tiliaceae	–	–	–	W	F, P

older floras (Table 14). Collinson (1983) detailed percentage climatic distribution of the nearest Recent analogues for British Tertiary floras. Early–early Middle Eocene floras contained the largest proportion of potential tropical (up to 92%) and the smallest potential temperate (up to 42%) elements. Late Palaeocene floras contained 70–78% tropical and 68–80% temperate elements. Overall, early Palaeogene fossil wood assemblages show a trend of increasing tropical elements through time: BIP flora 71% temperate, 14% tropical; Thanet Sand Formation 55%, 44%; Reading Formation 36%, 64%; Oldhaven Beds 23%, 65%; London Clay Formation 44%, 66%; overall Palaeocene/lowermost Eocene 51%, 44%; overall Lower Eocene 30%, 65%.

The Late Palaeocene environments of southern England consisted of flood plains, swamps and marshes, with more distant areas supporting types of vegetation such as those found at Newbury (Reading Formation). Woodland and forest surrounded the stable areas. Early Eocene vegetation was characterized by tropical forest (Collinson and Hooker 1987). The size of a woody remnant or a vulnerability index of 10+ for small remnants would indicate trees rather than shrubs (Wolfe and Upchurch 1987; Collinson and Hooker 1987; Wheeler 1991) (see Table 15). On the basis of these criteria the following were trees: BIP Flora, *Cercidiphylloxylon spenceri* (Cercidiphyllaceae); Thanet Sand Formation, *Anacardioxylon maidstonense* (Anacardiaceae, Burseraceae), *Cantia arborescens* (Betulaceae), *Castanoxylon philpii*

TABLE 14. Present day distribution of comparative families and genera grouped by formation (information from Metcalfe and Chalk 1950; Brazier and Franklin 1961; Witt 1966; and Collinson 1983).

Family	Genus	Present-day distribution
BRITO-ARCTIC IGNEOUS PROVINCE		
Cercidiphyllaceae	*Cercidiphyllum* spp.	Temperate to subtropical
Cunoniaceae	–	Temperate, Southern Hemisphere
Eucryphiaceae	–	Temperate, Southern Hemisphere
Eupteleaceae	*Euptelea* spp.	Subtropical, Asia
Fagaceae	*Castanea* spp.	Temperate to subtropical
Icacinaceae	–	Pantropical
Platanaceae	*Platanus* spp.	Temperate to subtropical
THANET SAND FORMATION		
Anacardiaceae	–	Temperate to tropical
Aquifoliaceae	–	Mainly temperate, some tropical
Betulaceae	–	Mainly temperate, some subtropical
Burseraceae	–	Pantropical
Euphorbiaceae	–	Pantropical, some temperate
Eupteleaceae	*Euptelea* spp.	Subtropical, Asia
Fagaceae	*Castanopsis* spp.	Tropical to subtropical, Asia
Icacinaceae	–	Pantropical
Platanaceae	*Platanus* spp.	Temperate to subtropical
READING FORMATION		
Anacardiaceae	–	Pantropical to subtropical
Burseraceae	–	Pantropical
Euphorbiaceae	–	Pantropical, some temperate
Eupteleaceae	*Euptelea* spp.	Subtropical, Asia
Fagaceae	*Fagus* spp.	Temperate to subtropical
Icacinaceae	–	Pantropical
Lauraceae	–	Pantropical to subtropical, some temperate
Meliaceae	*Entandrophragma* spp.	Tropical, Africa
Platanaceae	*Platanus* spp.	Temperate to subtropical
Sterculiaceae	*Tarrietia* spp.	Pantropical
OLDHAVEN BEDS		
Annonaceae	–	Pantropical
Apocynaceae	–	Pantropical to subtropical
Caesalpiniaceae	*Bussea* spp.	Tropical
Cercidiphyllaceae	*Cercidiphyllum* spp.	Temperate to subtropical
Cornaceae	*Curtisia* spp.	Subtropical, South Africa
Cunoniaceae	–	Tropical to subtropical, Southern Hemisphere
Eupteleaceae	*Euptelea* spp.	Subtropical, Asia
Flacourtiaceae	–	Pantropical to subtropical
Fagaceae	–	Subtropical to temperate, one genus pantropical
Hamamelidaceae	*Exbucklandia* spp., *Fothergilla* spp., *Hamamelis* spp., *Parriotiopsis* spp.	Subtropical to temperate Asia, and temperate North America
Icacinaceae	–	Pantropical
Lauraceae	*Cryptocarya* spp.	Pantropical
Lecythidaceae	–	Pantropical
Platanaceae	*Platanus* spp.	Temperate to subtropical
Sapotaceae	–	Pantropical
Theaceae	*Eurya* spp.	Tropical to subtropical Asia
Tiliacae	*Luehea* spp.	Tropical America

TABLE 14. *Continued*

Family	Genus	Present-day distribution
SUFFOLK PEBBLE BED		
Sterculiaceae	–	Pantropical
LONDON CLAY FORMATION		
Anacardiaceae	–	Pantropical to subtropical, some temperate
Celastraceae	–	Tropical to warm temperate
Cercidiphyllaceae	*Cercidiphyllum* sppp.	Temperate to subtropical
Dipterocarpaceae	*Anisoptera* spp.	Tropical, Malaysia
Eupteleaceae	*Euptelea* spp.	Subtropical, Asia
Fagaceae	*Lithocarpus* spp.	Pantropical
Icacinacese	–	Pantropical
Meliaceae	–	Pantropical to subtropical
Platanaceae	*Platanus* spp.	Temperate to subtropical
Rutaceae	*Fagara* spp.	Tropical, Africa
Sapindaceae	–	Mainly tropical, some temperate

(Fagaceae), *Edenoxylon aemulum* (Anacardiaceae), *Hythia elgarii* (Icacinaceae), *Quercinium porosum* (Fagaceae); Reading Formation, *Canarioxylon lewisii* (Anacardiaceae, Buseraceae), *Fagoxylon* sp. (Fagaceae), *Entandrophragminium lewisii* (Meliaceae), *Paraphyllanthoxylon lewisii* (Anacardiaceae, Burseraceae, Euphorbiaceae, Lauraceae), *Tarrietioxylon hazzeldinewarrenii* (Sterculiaceae); Oldhaven Beds, *Platanium decipiens* (Platanaceae?), *Tetrapleuroxylon oldhavenense* (Caesalpiniaceae). Boulter and Hubbard (1982) suggested that there were three major vegetational types in southern England during the Eocene: (1) temperate angiosperm deciduous forest; (2) subtropical fern and conifer forest; and (3) paratropical forest (climatic classification of Wolfe 1979). Major components of each are: (1) Betulaceae, Fagaceae, Hamamelidaceae; (2) Betulaceae, Fagaceae; and (3) Apocynaceae, Aquifoliaceae, Betulaceae, Bombacaceae, Icacinaceae, Sapotaceae, Tiliaceae. The wood data (Table 14) suggest that the vegetation during deposition of the Thanet Sand and London Clay formations was similar to types 2 and 3, whereas that of the Oldhaven beds was type 3.

Trends in xylem evolution

Features of these woods can be examined in the context of the major trends in xylem evolution (Bailey and Tupper 1918; Carlquist 1988).

1. Diffuse axial parenchyma is the most primitive, with more massive aggregations more specialized (Kribs 1937; Carlquist 1988). In the British woods both absent parenchyma or diffuse parenchyma characterize the BIP to Reading floras. By Oldhaven times a variety of specialized types (both in quantity and form) were present: banded, paratracheal aliform, confluent and combinations of these. The oldest records of banded and confluent parenchyma types are from Upper Cretaceous rocks (see Table 16).
2. Heterogeneous rays are considered primitive (Kribs 1935) and homocellular rays advanced (Carlquist 1988). The latter have not been recorded from British material below the Oldhaven Beds (*Annonoxylon* and *Tilioxylon*).
3. Storied wood structure is associated with short fusiform initials and correlates with shortening vessel-element length as a major evolutionary trend (Carlquist 1988). No British woods show regular storied structure, although locally storied axial parenchyma (*Tarrietioxylon hazzeldinewarrenii* Crawley, 1989) or locally storied rays or rays en echelon are seen (*Edenoxylon aemulum* and *Tarrietioxylon hazzeldine-warrenii*). No storied structure has been described from wood older than Palaeogene (see Table 16).
4. Solitary vessels with tracheids, or radial groups with fire-tracheids or libriform fibres are characters related to dryness of habitat (Carlquist 1988). Solitary vessels are present in the BIP flora, and the floras of the Thanet Sand and London Clay formations. Radial groups are present in all but the BIP flora, but groups

TABLE 15. Features of wood anatomy with climatic and ecological significance from the British Palaeogene. SPP, scalariform perforation plates; GR, growth rings; VI, vulnerability indices; +, present; (+) present but indistinct.

	SPP	GR	VI
LOWER GREENSAND			
Aptiana radiata Stopes, 1912	–	(+)	0·4
BRITO-ARCTIC IGNEOUS PROVINCE			
Cercidiphylloxylon spenceri (Brett) Pearson, 1997	+	+	0·3
Eiggoxylon reidii Crawley, 1989	+	+	1·9
Plataninium brettii Crawley, 1989	+	+	0·7
THANET SAND FORMATION			
Anacardioxylon maidstonense sp. nov.	–	(+)	5·1, 1·0
Cantia arborescens Stopes, 1915	+	(+)	0·35
Castanoxylon maidstonense sp. nov.	–	+	20
Edenoxylon aemulum Brett, 1966 (this study)	–	+	2·6
Edenoxylon aemulum Brett, 1966	–	+	2·2
Euphorbioxylon hernense sp. nov.	–	(+)	0·5
Hythia elgari Stopes, 1915	+	–	0·4
Ilicoxylon? *prestwichii* sp. nov.	+	+	–
Quercinium porosum Brett, 1972	+	(+)	2·7
Ulminium elliottii Crawley, 1989	–	+	15·5
READING FORMATION			
Canarioxylon lewisii sp. nov.	–	–	6·3
Fagoxylon sp.	+	(+)	0·4
Entandrophragminium lewisii sp. nov.	–	–	35
Paraphyllanthoxylon lewisii sp. nov.	–	–	14·5
Plataninium decipiens Brett, 1972, *in* Crawley, 1989	+	–	4·5
Tarrietioxylon hazeldinewarrenii Crawley, 1989	–	–	39·2
Tarrietioxylon cf. *hazeldinewarrenii* Crawley, 1989	–	–	25
OLDHAVEN BEDS			
Apocynoxylon oldhavenense sp. nov.	–	(+)	7
Cercidiphylloxylon cf. *kadanense* Prakash *et al*, 1971	+	+	1·7
Sabulia scottii Stopes, 1912	+	(+)	6·25
Dryoxylon calodendrumoides sp. nov.	–	–	2·4
Flacourtioxylon oldhavenense sp. nov.	–	(+)	1·2
Hamamelidoxylon renaultii Lignier, 1907	+	+	0·2
Plumerioxylon ipswichense gen. et sp. nov.	–	(+)	4
Polyalthioxylon oldhavenense sp. nov.	–	(+)	3·9
Tetrapleuroxylon oldhavenense sp. nov.	–	(+)	13·3
Tilioxylon lueheaformis sp. nov.	–	(+)	0·5
SUFFOLK PEBBLE BED			
Dombeyoxylon cf. *sturanii* Charrier, 1967	–	(+)	15
LONDON CLAY FORMATION			
Cercidiphylloxylon spenceri (Brett) Pearson, 1987	+	+	0·25
Fagaroxylon atkinsoniae (Crawley) comb. nov.	–	(+)	3·5
Meliaceoxylon collinsoniae sp. nov.	–	–	12
Quercinium pasanoides Brett, 1960	–	(+)	14·4
Plataninium decipiens Brett, 1972	+	(+)	2·1

of 4+ only by Oldhaven times (the earliest occurrences are Late Cretaceous, Table 16). When these British woods are compared to characteristics previously reported for the Palaeocene woods (Wheeler and Baas 1991) the following features are of note: (1) the second record of en echelon ray structure, present in *Edenoxylon aemulum* (the first was noted in *Caesalpinioxylon moragjonesae* Crawley, 1988, from the Palaeocene of the Republic of Mali); (2) the first record of tangential vessel arrangement present in

TABLE 16. Significant features of wood anatomy and their earliest geological occurrences.

SEMI-RING POROUS
Eiggoxylon reidii Crawley, 1989 (Cunoniaceae?, Eucryphiaceae?), Palaeocene
Paraquercinium cretaceum Wheeler, Lee and Matten, 1987 (Fagaceae), Upper Cretaceous
Riboidoxylon cretacea Page, 1967 (Escalloniaceae?, Aquifoliaceae?), Upper Cretaceous
AXIAL PARENCHYMA STORIED
Tarrietioxylon hazeldinewarrenii Crawley, 1989, Palaeocene/lowermost Eocene
RAYS STORIED
Javelinioxylon multiporosum Wheeler, Lehman and Gasson, 1994, Upper Cretaceous
VESSELS IN RADIAL OR OBLIQUE ARRANGEMENT
Carpinoxylon ostryopsoides Page, 1967 (Betulaceae), CASG 60428 Page, 1980, Upper Cretaceous
Euphorbioxylon bussonii Koeniguer, 1968 (Euphorbiaceae), Upper Cretaceous
Mulleroxylon eupomatioides Page, 1967 (Eupomatiaceae), Upper Cretaceous
VESSELS IN TANGENTIAL BANDS
Plataninium brettii Crawley, 1989 (Icacinaceae), Palaeocene
cf. *Plataninium haydenji* Wheeler, 1991 (Platanaceae?), Palaeocene
VESSEL MULTIPLES OF > 4 COMMON
Dicotyloxylon sp. 1 Torres and Lemoigne, 1989, Upper Cretaceous
Laurinoxylon weylandii Berger, 1953 (Lauraceae), Upper Cretaceous
Paraphyllanthoxylon illinoisense Wheeler, Lee and Matten, 1987 (Anacardiaceae?, Buseraceae?, Euphorbiaceae?), Upper Cretaceous
AXIAL PARENCHYMA CONFLUENT
Cassinium dongolense Giraud and Lejal-Nicol, 1989 (Caesalpiniaceae), Upper Cretaceous
AXIAL PARENCHYMA BANDED
Cassinium dongolense Giraud and Lejal-Nicol, 1989 (Caesalpiniaceae), Upper Cretaceous
Euphorbioxylon nigerinum Koeniguer, 1971 (Euphorbiaceae), Lower Palaeocene
Myristicoxylon princeps Boureau, 1950*b* (Myristicaceae), Lower Palaeocene
Parapocynaceoxylon barghoorni Wheeler, Lee and Matten, 1987 (Apocynaceae?), Upper Cretaceous
Paratrichilioxylon? *russellii* Koeniguer, 1971 (Meliaceae), Lower Palaeocene
RAYS UNISERIATE
CASG 60117 Page, 1979, Upper Cretaceous
CASG 60426 Page, 1980, Upper Cretaceous
Paraquercinium cretaceum Wheeler, Lee & Matten, 1987 (Fagaceae), Upper Cretaceous
RAYS HOMOCELLULAR
Carpinoxylon ostryopsoides Page, 1967 (Betulaceae), Upper Cretaceous
CASG 60130 Page, 1979, Upper Cretaceous
CASG 60208 Page, 1979, Upper Cretaceous
CASG 60428 Page, 1980, Upper Cretaceous
CASG 60447 Page, 1980, Upper Cretaceous
Cassinium dongolense Giraud and Lejal-Nicol, 1989 (Caesalpiniaceae), Upper Cretaceous
Eiggoxylon reidii Crawley, 1989, Palaeocene (Cunoniaceae?, Eucryphiaceae?), Upper Cretaceous
Hedycaryoxylon affine (Vater) Süss, 1960 (Monimiaceae), Upper Cretaceous
Lardizabaloxylon cocculoides Page, 1970, Upper Cretaceous
Magnoliaceoxylon (*Magnolioxylon*) *panochensis* (Page, 1967) Wheeler, Scott & Barghoorn, 1977 (Magnoliaceae), Upper Cretaceous
Mimosoxylon tenax (Felix) Müller-Stoll and Mädel, 1967 (Mimosaceae?), Upper Cretaceous
Nothofagoxylon scalariforme Gothan, 1908; Nishida, Nishida and Rancusi, 1988 (Fagaceae), Upper Cretaceous
Parapocynaceoxylon barghoorni Wheeler, Lee and Matten, 1987 (Apocynaceae?), Upper Cretaceous
Paraquercinium cretaceum Wheeler, Lee and Matten, 1987 (Fagaceae?), Upper Cretaceous
Plataninium californicum Page, 1968 (Platanaceae?), Upper Cretaceous
cf. *Plataninium haydenii* Wheeler, 1991 (Platanaceae?), Upper Cretaceous
Plataninium platanoides Page, 1968 (Platanaceae?), Upper Cretaceous

Dryoxylon calodendrumoides; (3) the first record of helical thickenings in vessel elements, present in *Tilioxylon lueheaformis*; (4) the first record of radial canals, present in *Edenoxylon aemulum*; (5) The second and third records of axial canals, present in *Entandrophragminium lewisii* and *Ilicoxylon*? *prestwichii* (*Caesalpinioxylon moragjonesae* Crawley, 1988 was the first).

Scott and Wheeler (1982) and Wheeler and Baas (1991, 1993) have shown that in the Northern Hemisphere the proportion of woods with scalariform perforation plates decreases from the Cretaceous Period onwards; Cretaceous, 60%; Eocene 35%; Miocene 25%. In this study, the proportion of woods with scalariform plates are: BIP flora, 100%; Thanet Sand Formation, 36%; Reading Formation, 28%; Oldhaven Beds, 25%; Suffolk Pebble Beds, 0%; London Clay Formation, 40%. The average for all the British Palaeocene/lowermost Eocene woods is 41%. The overall trend supports the reduction of woods with scalariform perforation plates through time.

Of interest is *Aptiana* which, although exhibiting some primitive features *sensu* Bailey, also has alternate minute and unilaterally compound pitting. Carlquist (1988) regarded as uncertain the relationship of minute pitting to salient lines of wood evolution. Unilaterally compound pitting has not been evaluated. That this pit type occurs in the fossil record before the scalariform type is noted here, but with such small samples it is unwise to draw any conclusion from this.

Palaeoclimatic interpretation

Hitherto, generalized Palaeocene climatic inferences from wood remains have been considered preliminary at best owing to the lack of material (Wheeler and Baas 1991). The 23 new Palaeocene records described herein allow some re-evaluation of this position (see Table 13 for ecologically significant data and Table 15 for present-day distribution of families and genera).

Growth rings, particularly ring porosity, are taken as evidence of periodicity of climate (Gilbert 1940). Reviewing these data, the woods of the BIP flora have distinct growth rings, and one, *Eiggoxylon*, has a semi-ring porous structure. Both suggest a seasonal environment (Crawley 1989). However, the trend elsewhere is of indistinct or a lack of growth rings up to Early Eocene times, indicating climates in which there was little or no seasonality. A survey by Wheeler and Baas (1991) yielded similar results, which led them to conclude that the fairly 'low' percentage (36%) of woods with distinct growth rings indicates a largely tropical flora. Reports of 'high' incidences of growth rings (31%) in the angiosperm twig flora of the London Clay Formation (Scott and de Klerk 1974; Collinson 1983) are not corroborated by the log material (Crawley 1989 and herein) or in recent studies of the twigs (Poole 1992, 1999), in which growth rings are usually indistinct or lacking. Growth ring data from woods of the London Clay Formation need careful analysis as today the number of tropical tree species forming annual rings is much greater than previously assumed (Boninsegna *et al.* 1989; Worbes 1989) and can vary according to genera rather than climate (Détienne 1989).

High frequencies of helical thickening (39–42%; Wheeler and Baas 1991) occur in wood from dry, Mediterranean-type areas (Carlquist 1988); low frequencies occur in tropical floras (2–5%; Wheeler and Baas 1991). Helical thickenings are present in the vessels of *Tilioxylon ipswichense*, the earliest occurrence in the fossil record to date. However, the value of this feature for ecological purposes cannot be considered reliable below the mid to upper Tertiary (Wheeler and Baas 1991).

Species with vasicentric tracheids such as *Castanoxylon philpii, Quercinium porosum* (both Thanet Sand Formation) and *Quercinium pasanoides* (London Clay Formation) occur in Mediterranean-type areas (warm throughout year, dry summer, moderately wet winter; Carlquist 1988) as well as subtropical–tropical areas (E. A. Wheeler, pers comm. 1991).

According to Carlquist (1977), vulnerability or V indices (mean vessel diameter divided by vessels mm^2) can also yield ecological data. Both Wheeler (1991) and Poole (1994) suggested that it is necessary to be cautious with fossil material. Correlation of fossil twig values with data from modern examples requires greater precision for palaeoecological interpretation (Poole 1994). The validity of V indices in Recent wood may also be questioned. Wheeler (1991) has, however, noted that high V indices may be of use in a general way as a characteristic of groups of Recent tropical and pioneer trees. She also stated in defense of V values that trees with diffuse porous wood and large diameter vessels do not occur in, for example, temperate regions of pronounced seasonality. Carlquist's mean V value for primitive wood growing in an environment of equable water stress (mesophytic) is 2·29. As noted previously (Crawley 1989) all the BIP woods fall below this value and have small diameter vessels and some ring porosity. This is indicative of high water stress and a distinctly seasonal, temperate climate. The woods of the Thanet Sand Formation, Oldhaven Beds and London Clay Formation (see Table 15) are around Carlquist's mesophytic value but some diffuse porous woods with large diameter vessels also occur, pointing to low water stress and a non-temperate climate. The Oldhaven specimens are biased towards small axis material. In twigs and small branches of Recent wood axial elements are smaller, both axially and radially, than elements in trunks, large branches or roots; small axes have higher V values than large axes (Barefoot and Hankins 1982). The solitary Oldhaven wood with a high V index is from a larger axis. Most woods from the Reading Formation and Suffolk Pebble Beds have higher values and are diffuse porous with usually medium to large vessels, indicating low water stress or even a stress-free environment and an equable, non-temperate climate. Recently Wiemann *et al.* (1999) have used some dicotyledonous wood characters for estimating palaeotemperature based on earlier findings with Recent woods (Wiemann *et al.* 1998). They concluded that a sample of at least 20 species from a single habitat should be used if a worthwhile conclusion is to be drawn; hence, this type of study has not been attempted here.

From this variety of data it seems that the BIP flora grew in distinctly seasonal conditions of temperate (Seward and Holltum 1924) to subtropical (Boulter and Kvaček 1989) type. Less marked seasonality and a warmer temperature is indicated by Thanet times with an apparently aseasonal, probably tropical to subtropical climate during Reading times. A similarly warm climate with perhaps some seasonality is suggested by the woods of the Oldhaven Beds and London Clay formation. Daley (1972) proposed a frost-free climate for the early Palaeogene of southern England. Oxygen isotope work by Buchardt (1978) yielded a mean annual palaeotemperature of between 16 and 25°C (late Thanetian) and 20–27°C (early Eocene) in the North Sea area. A hot climate was also supported by Shackleton (1986) who recorded a deep ocean palaeotemperature of 10°C for the Palaeocene, with maximum of 12°C occurring in Early or early Mid Eocene. These values are equivalent to the present-day subtropical/paratropical forest classification of Wolfe (1979). Hubbard and Boulter (1983) suggested an unstable but equable climate in north-west Europe during the Late Palaeocene–Early Eocene; frostless with summer maxima of 21–25°C. If not evidence of an unstable climate, at least variety of climate may be indicated by the distinct wood assemblages from the Thanet, Reading and Oldhaven beds. Collinson and Hooker (1987) suggested that the 'tropicality' of the Early Eocene may have been under emphasized, citing bias in the fossil record towards 'edge effect' floral components, such as occur at the borders of tropical rainforests and tending to occur along stream-sides. These 'edge effect' floral components include elements normally associated with more temperate vegetation.

Provenance of ex situ *specimens*

Petrified, particularly silicified, woods are amazingly resilient and can easily be transported great distances without any obvious wear and tear. A cautious approach to their study is, therefore, always advisable. The following macro-features in fossil wood are, in my opinion, key to permitting association or grouping of particular specimens not only with each other but also with particular geological horizons when specimens are found *ex situ*: general colour, both externally and internally; external features such as signs of weathering; presence or lack of adherent matrix; preservational type (calcified, silicified).

Useful information can often be gained from any adherent matrix or, as for the fossils studied here, from burrow fillings. Curators/collectors should not clean up any fossil wood specimens prior to incorporation into a collection because this can remove vital evidence of provenance. Also I would emphasize the importance of good curation. The specific value of any specimen is dependent upon the associated information which makes that specimen scientifically valuable. This information needs to be recorded and stored with as much care as the specimen itself.

Acknowledgements. Most of this study was completed while I was a member of the curatorial staff of the Department of Palaentology, The Natural History Museum, London. This work has spanned the period from 1989 to date, and many people have been involved along the way. I thank them all for their assistance. In particular, thanks are due to Dr L. R. M. Cocks (now retired), Keeper of Palaeontology, NHM; Mr R. A. D. Markham, Curator of Geology, Ipswich Museum; Mr E. G. Philp (now retired), Maidstone Museum and Art Gallery; and Mr A. Higgott, Curator of Newbury Museum, for allowing access to collections in their care; Prof. V. M. Page, Stanford University, California, for her valuable assistance regarding *Dipterocarpoxylon porosum*; Dr D. Cutler and Dr P. Gasson, Jodrell Laboratory, Royal Botanic gardens, Kew, for allowing study of the slide collection of Recent woods; Prof. E. A. Wheeler, Department of Wood and Paper Science, North Carolina State University, for invaluable discussion and encouragement; Mr A. J. H. Wighton, for his skill in producing the many thin sections examined; Mr D. Moore and Mr H. Buckley, Department of Mineralogy, NHM, for analysis of a variety of mineralized infillings; Mr M. Beasley and Mrs A. Lum, Botany and Palaeontology librarians, NHM; Mr L. R. Lewis, who collected much of the material from the Reading Formation and first brought it to my attention, and for his hospitality when I visited the locality; the late Dr G. F. Elliott, Dr C. R. Hill and Mr C. H. Shute, NHM, for their support and helpful suggestions with the manuscript; Prof. M. C. Boulter and Mr J. Cocking, University of East London who acted as supervisors and mentors for this project, part of my submission for an MSc degree; members of the Central Photographic Processing Unit, NHM; Mr J. Cooper, Dr J. J. Hooker, Dr H. G. Owen, Mr D. Ward and Dr H. P. Wilkinson for valuable discussion; Dr I. Poole for providing much help with recent publications; and Ms Pat Kerr for her reliable support during the period of this study.

REFERENCES

AWASTHI, N. 1970. On the occurrence of two new fossil woods belonging to the family Lecythidaceae in the Tertiary rocks of South India. *Palaeobotanist*, **18**, 67–74.

AUBRÉVILLE, A. 1961. Notes sur des Poutérieés Américaines. *Adansonia*, **1**, 150–191.

BAAS, P. 1973. The wood anatomical range in *Ilex* (Aquifoliaceae) and its ecological and phylogenetic significance. *Blumea*, **21**, 193–258.

—— 1975. Vegetative anatomy and the affinities of Aquifoliaceae, *Sphenostemon, Phelline* and *Oncotheca. Blumea*, **22**, 311–407.

BAILEY, I. W. 1924. The problem of identifying the wood of Cretaceous and later dicotyledons, *Paraphyllanthoxylon arizonense*. *Annals of Botany, London*, **38**, 439–452.

—— and TUPPER, W. W. 1918. Size variation in tracheary cells. 1. A comparison between secondary xylems of vascular cryptograms, gymnosperms and angiosperms. *Proceedings of the American Academy of Arts and Sciences*, **54**, 149–204.

BALSON, P. S. 1980. The origin and evolution of Tertiary phosphorites from eastern England. *Journal of the Geological Society, London*, **137**, 723–729.

BANCROFT, H. 1932. On the identification of isolated timber specimens, with especial reference to fossil woods. *Annals of Botany, London*, **182**, 353–365.

—— 1933. A contribution to the geological history of the Dipterocarpaceae. *Geologiska Föreningens i Stockholm Förhandlingar*, **55**, 59–100.

BANDE, M. B. 1973. A petrified dicotyledonous wood from the Deccan Intertrappean beds of Mandla District, Madhya Pradesh. *Botanique*, **4**, 41–48.

—— 1974. Two fossil woods from the Deccan Intertrappean beds of Mandla District, Madhya Pradesh. *Geophytology*, **4**, 189–195.

—— and KHATRI, S. K. 1980. Some more fossil woods from the Deccan Intertrappean beds of Mandla District, Madhya Pradesh, India. *Palaeontographica, Abteilung B*, **173**, 147–165.

—— and PRAKASH, U. 1983. Fossil dicotyledonous wood from the Deccan Intertrappean beds near Shahpura, Mandla District, Madhya Pradesh. *Palaeobotanist*, **31**, 13–29.

BAREFOOT, A. C. and HANKINS, F. W. 1982. *Identification of modern and Tertiary woods*. Clarendon Press, Oxford, vii + 189 pp.

BARTLING, F. G. 1830. *Ordines naturales Plantarum: adjecta generum enumeratione*. Gottingae, iv + 498 pp.

BERGER, W. 1953. Ein Lauraceenholz aus dem Oberkreideflysch des Lainzer Tiergartens bei Wien: *Laurinoxylon weylandi* n. sp. *Osterreich Botanisch Zeitschrift*, **100**, 136–146.

BIERHORST, D. W. 1960. Observations on tracheary elements. *Phytomorphology*, **10**, 249–305.

BLISS, M. C. 1921. The vessel in seed plants. *Botanical Gazette*, **71**, 314–326.

BLOKHINA, N. I. and SNEZHOVA, S. A. 1999. *Alnus tschemrylica* sp. nov. (the Betulaceae) from the Palaeogene of Tschermunaut bay (Kamchatka). *Palaeontological Journal*, **33**, 131–136.

BLUME, K. L. 1825. *Bijdragen tot de Flora van Nederlandsch Indie.* (2 volumes). Batavia, 1169 pp.

BONINSEGNA, J. A., VILLALBA, R., AMARILLA, L. and OCAMPO, J. 1989. Studies on tree rings, growth rates and age-size relationships of tropical tree species in Misione, Argentina. *Bulletin of the International Association of Wood Anatomists, New Series*, **10**, 161–169.

BOULTER, M. C. and HUBBARD, R. N. L. B. 1982. Objective paleoecological and biostratigraphic interpretation of Tertiary palynological data by multivariate analysis. *Palynology*, **6**, 55–68.

—— and KVAČEK, Z. 1989. The Palaeocene flora of the Isle of Mull. *Special Papers in Palaeontology*, **42**, 149 pp.

—— and MANUM, S. B. 1989. The 'Brito-Arctic Igneous Province' flora around the Palaeocene/Eocene boundary. *Proceedings of the Ocean Drilling Program*, **104B**, 663–680.

BOUREAU, E. 1950*a*. Étude paleoxylologique du Sahara. 12. Sur un *Annonoxylon striatum* n. gen. et n. sp. des couches de Tamaguilel (Sahara soudanais). *Bulletin de la Société Géologique de France, Séries 5*, **20**, 393–397.

—— 1950*b*. Étude paleoxylologique du Sahara. 9. Sur un *Myristicoxylon princeps* n. gen., n. sp., du Danien d'Asselar (Sahara soudanais). *Bulletin du Muséum d'Histoire Naturelle, Paris, Deuxieme Série*, **22**, 523–528.

—— 1954. Étude paleoxylologique du Sahara. 20. Sur un *Annonoxylon edengense* n. sp. des couches post-eocene du Sud-Ouest de l'Adrar Tiguirit (Sahara soudanais). *Bulletin du Muséum d'Histoire Naturelle, Paris, Deuxieme Série*, **26**, 286–291.

BOWERBANK, J. S. 1840. *A history of the fossil fruits and seeds of the London Clay.* John Van Voorst, London,144 pp.

BRAZIER, J. D. and FRANKLIN, G. L. 1961. Identification of hardwoods: a microscope key. *Forest Products Research Bulletin*, **46**, 1–96.

BRETT, D. W. 1956. Fossil wood of *Cercidiphyllum* Sieb. and Zucc. from the London Clay. *Annals and Magazine of Natural History; including Zoology, Botany and Geology, Series 12*, **9**, 657–665.

—— 1960. Fossil oak wood from the British Eocene. *Palaeontology*, **3**, 86–92.

—— 1966. Fossil wood of Anacardiaceae from the British Eocene. *Palaeontology*, **9**, 360–364.

—— 1972. Fossil wood of *Platanus* from the British Eocene. *Palaeontology*, **15**, 496–500.

BUCHARDT, B. 1978. Oxygen isotope palaeotemperatures from the Tertiary period in the North Sea area. *Nature*, **275**, 121–123.

BUNGE, A. 1851. Beitrag zur Kenntniss der Flora Russlands und der Steppen central-Asiens. *Mémoires Présentés à l'Académie Impériale des Science de St. Petersburg par divers Savants et dans ses Assemblées*, **7**, 177–536.

BURGH, J. VAN DER 1964. Hölzer der niederrheinischen Braunkohlenformation 1. *Acta Botanica Neerlandica*, **15**, 276–283.

—— 1973. Hölzer der Braunkohlenformation, 2. Hölzer der Braunkohlengruben 'Maria Theresia' zu Herzogenrath, 'Zukunft West' zu Eschweiler und 'Victor' (zülpich mitte) zu Zülpich. Nebst einer systematisch-anatomischen Bearbeitung der gattung *Pinus* L. *Review of Palaeobotany and Palynology*, **15**, 73–275.

—— 1978. Hölzer aus dem Pliozän der Niederrheinischen Bucht. *Fortschritte in der Geologie von Rheinland und Westfalen*, **28**, 213–275.

CAHOON, E. J. 1972. *Paraphyllanthoxylon albamense*, a new species of fossil dicot wood. *American Journal of Botany*, **59**, 5–11.

CANDOLLE, A. P. de 1824. *Prodromus systematis naturalis regni vegetabilis.* Paris, 747 pp.

CARLQUIST, S. 1977. Ecological factors in wood evolution, a floristic approach. *American Journal of Botany*, **64**, 887–896.

—— 1988. *Comparative wood anatomy*, Springer-Verlag, Berlin, xi + 354 pp.

—— 1989. Wood and bark anatomy of the New World species of *Ephedra. Aliso*, **12**, 441–483.

CHAMBERLAIN, C. J. 1935. *Gymnosperm structure and evolution.* Chicago, viii + 484 pp.

CHANDLER, M. E. J. 1961. *The lower Tertiary floras of southern England.* **1**. *Palaeocene floras, London Clay flora* (supplement). Text and atlas. The Natural History Museum, London, xi + 354 pp.

—— 1963. *The lower Tertiary floras of southern England.* **3**. *Flora of the Bournemouth Beds; the Boscombe and the Highcliff sands.* The Natural History Museum, London, xi + 169 pp.

—— 1964. *The lower Tertiary floras of southern England.* **4**. *A summary and survey of findings in the light of recent botanical observations.* The Natural History Museum, London, xi + 150 pp.

—— 1978. Supplement to the Lower Tertiary floras of southern England. 5. *Tertiary Research, Special Paper*, **4**. 47 pp.

CHARRIER, G. 1967. Legno di Sterculiaceae dell'eocene medio continentale del Lauzanier (autoctone sedimentario dell'Argentera, Basses Alpes, Francia). *Bollettino della Società Geologica Italiana*, **86**, 733–747.

COLLINSON, M. E. 1983. *Fossil plants of the London Clay.* Palaeontological Association, Field Guides to Fossils, **1**. The Palaeontological Association, London, 121 pp.

—— and CRANE, P. R. 1978. *Rhododendron* seeds from the Palaeocene of southern England. *Botanical Journal of the Linnaean Society*, **76**, 195–205.

—— and HOOKER, J. J. 1987. Vegetational and mammalian faunal changes in the early Tertiary of southern England. 259–304. *In* FRIIS, E. M., CHALONER, W. G. and CRANE, P. R. (eds). *The origin of angiosperms and their biological consequences*. Cambridge University Press, Cambridge, 358 pp.

—— and RIBBINS, M. M. 1977. Pyritized fern rachides in the London Clay. *Tertiary Research*, **1**, 109–113.

CRANE, P. R. 1978. Angiosperm leaves from the Lower Tertiary of southern England. *Courier Forschungsinstitut Senckenberg*, **30**, 126–132.

—— 1981. Betulaceous leaves and fruits from the British Upper Palaeocene. *Botanical Journal of the Linnaean Society*, **83**, 103–136.

—— 1984. A re-evaluation of *Cercidiphyllum*-like plant fossils from the British early Tertiary. *Botanical Journal of the Linnean Society*, **89**, 199–230.

—— 1988. Major clades and relationships in the 'higher' gymnosperms. 218–272. *In* BECK, C. B. (ed.). *Origin and evolution of gymnosperms*. Columbia University Press, New York, xiv + 504 pp.

—— and GOLDRING, R. 1991. The Woolwich and Reading Formation (late Palaeocene–early Eocene) at Cold Ash and Pincent's Kiln (Berks) in the western London Basin. *Tertiary Research*, **12**, 147–158.

—— and MANCHESTER, S. R. 1982. An extinct juglandaceous fruit from the Upper Palaeocene of southern England. *Botanical Journal of the Linnean Society*, **85**, 89–101.

—— and STOCKEY, R. A. 1987. *Betula* leaves and reproductive structures from the Middle Eocene of British Colombia, Canada. *Canadian Journal of Botany*, **65**, 2490–2500.

CRAWLEY, M. 1988. Palaeocene wood from the Republic of Mali. *Bulletin of the British Museum (Natural History), Geology*, **44**, 3–14.

—— 1989. Dicotyledonous wood from the lower Tertiary of Britain. *Palaeontology*, **32**, 597–622.

CREBER, G. T. and CHALONER, W. G. 1984. Climatic indications from growth rings in fossil woods. 49–74. *In* BRENCHLY, P. (ed.). *Fossils and climate*. John Wiley and Sons, Chichester, xi + 352 pp.

—— and FRANCIS, J. E. 1999. Fossil tree-ring analysis: palaeodendrology, 245–250. *In* JONES, T. P. and ROWE, N. P. (eds). *Fossil plants and spores: modern techniques*. Geological Society, London, xii + 396 pp.

CRONQUIST, A., TAKHTAJAN, A. and ZIMMERMAN, W. 1966. On the higher taxa of Embryobionta. *Taxon*, **15**, 129–134.

DALEY, B. 1972. Some problems concerning the early Tertiary climate of southern Britain. *Palaeogeography, Palaeoclimatology, Palaeoecology*, **11**, 177–190.

DELTEIL-DESNEUX, F. and FESSLER-VROLANT, C. 1976. Sur la présence d'un bois de Légumineuse dans l'Oligocène de la Tunisie septentrionale. *Compte Rendu du Congrés des Sociétés Savantes de Paris et des Départments, Section des Sciences, Paris*, **101e** (Lille, 1), 185–195.

DEN BERGER, L. G. 1927. Unterscheidungsmerkmale von rezentenund fossilen Dipterocarpaceengattungen. *Bulletin du Jardin Botanique de Buitenzorg, Series 3*, **8**, 495–498.

DÉTIENNE, P. 1989. Appearance and periodicity of growth rings in some tropical woods. *Bulletin of the International Association of Wood Anatomists, New Series*, **10**, 123–132.

—— and JACQUET, P. 1983. *Atlas d'identification des bois l'Amazonie et des régions voisines*. Centre Technique Forestier Tropical, Nogent-sur-Marne, 640 pp.

DINES, H. G. and EDMUNDS, F. H. 1933. *The geology of the country around Reigate and Dorking*. Memoirs of the Geological Survey of England and Wales, **286**. HMSO, London, vii + 204 pp.

DUMORTIER, B. C. J., COUNT 1829. *Analyse des familles des plantes, &c.* Tournay, 104 pp.

DUPÉRON, J. 1973. *Oleoxylon aginnense* n. gen., n. sp., *Pistacioxylon muticoides* n. gen., n. sp. bois fossiles tertiaires de la région de Grateloup (Lot-et-Garonne). *Bulletin de la Société Botanique de France*, **120**, 311–330.

—— 1975. *Tetrapleuroxylon aquitanense* n. sp. bois fossile tertiare de la Molasse de l'Agenais. *Compte Rendu du Congrés des Sociétés Savantes de Paris et des Départments. Section des Sciences, Paris*, **100e** (Paris, 2), 63–73.

DUPÉRON-LAUDOUENEIX, M. 1977. Les bois heteroxyles de la region de Tibati (Cameroun). 1. Étude d'une structure de Flacourtiaceae. *Compte Rendu du Congrés des Sociétés Savantes de Paris et des Départments. Section des Sciences, Paris*, **102e** (Limoges, 1), 127–141.

EDWARDS, W. N. 1931. Dictyledones (Ligna). *Fossilium Catalogus* **2**, *Plantae*, **17**, 96 pp.

ELLISON, R. A., KNOX, R. W. O'B., JOLLEY, D. W. and KING, C. 1994. A revision of the lithostratigraphical classification of the early Palaeogene strata of the London Basin and East Anglia. *Proceedings of the Geologists' Association*, **105**, 187–197.

ENGLER, H. G. A. 1909. *Syllabus der Pflanzenfamilien*. Berlin, xxviii + 254 pp.

FELIX, J. 1882. *Studien über Fossile Hölzer*. Inaugural Dissertation. Leipzig, 79 pp.

—— 1883. Untersuchungen über fossile Hölzer. 1. *Zeitschrift der Deutschen Geologischen Gesellschaft*, **35**, 59–91.

—— 1884. Die Holzopale Ungarns in paläophytologischer Hinsict. *Mitteilungen aus dem Jahrbuch der Könlich Ungarishen Geologischen Anstalt*, **7**, 1–43.

—— 1887. Untersuchungen über fossile Hölzer. 3. *Zeitschrift der Deutschen Geologischen Gesellschaft*, **39**, 517–528.

FLINDERS, M. 1814. *A voyage to Terra Australis in 1801–1803 in H.M.S. The Investigator*. London, 10 pls, 18 maps.

GAZEAU-KOENIGUER, F. 1975. Étude d'un bois heteroxyle de l'Eocène du Vexin français. *Compte Rendu du Congrés des Sociétés Savantes de Paris et des Départments. Section des Sciences, Paris*, **100e** (Paris, 2), 75–81.

GHOSH, P. K. and ROY, S. K. 1978. Fossil wood of *Canarium* from the Tertiary of West Bengal, India. *Current Science*, **47**, 804–805.

—— and ROY, S. K. 1979. Fossil wood of *Dracontomelum* from the Tertiary of West Bengal. *Current Science*, **48**, 362.

GILBERT, S. G. 1940. Evolutionary significance of ring porosity in woody angiosperms. *Botanical Gazette*, **102**, 105–120.

GIRAUD, B. and LEJAL-NICOL, A. 1989. *Cassinium dongolense* n. sp. bois fossile de Caesalpiniaceae du Nubien du Soudan Septentrional. *Review of Palaeobotany and Palynology*, **59**, 37–50.

GOTHAN, W. 1908. Die fossilen Hölzer von den Seymour und Snow Hill Inseln. *Wissenschaftliche Ergebnisse der Schwedischen Südpolar–Expedition*, 1901–1903, **3**, 1–33.

GOTTWALD, H. 1969. Zwei Kieselhölzer aus dem Oligozan von Tunis *Bombacoxylon oweni* und *Pseudolachonostyloxylon weylandii*. *Palaeontographica, Abteilung B*, **125**, 112–118.

GRAMBAST-FESSARD, N. 1969. Contribution à des flores tertiaires des regions Provencales et alpines, 5. Deux bois de dicotylédones à caractères primitifs du Miocène. *Naturalia Monspeliensia, Séries Botanique*, **20**, 105–118.

GRAY, S. F. 1821. *A natural arrangement of British plants ... including those cultivated for use with an introduction to botany*. (2 volumes), London, 250 pp.

GREGUSS, P. 1969. *Tertiary angiosperm woods in Hungary*. Akadémiai Kiadó, Budapest, 151 pp.

HALL, J. W. 1952. The comparative anatomy and phylogeny of the Betulaceae. *Botanical Gazette*, **113**, 235–270.

HARLAND, W. B., ARMSTRONG, R. L., COX, A. V., CRAIG, L. E., SMITH, A. G. and SMITH, D. G. 1990. *A geologic time scale 1989*. Cambridge University Press, Cambridge, xvi + 263 pp.

HEIMSCH, C. 1942. Comparative anatomy of the secondary xylem in the Gruinales and Terebinthales, of Wettstein with reference to taxonomic grouping. *Lilloa*, **8**, 83–198.

HERENDEEN, P. S. and CRANE, P. R. 1992. Early caesalpinoid fruits from the Palaeocene of southern England. 57–68. *In* HERENDEEN, P. S. and DILCHER, D. L. (eds). *Advances in legume systematics 4. The fossil record.* Royal Botanic Gardens, Kew, 326 pp.

HILLEBRAND, W. 1888. *Flora of the Hawaiian Islands; a description of their phanerogams and vascular cryptograms.* London, cxvi + 673 pp.

HOFMANN, E. 1934. Ein fossiles Holz aus dem Rhyolith-Tuff des Tokayer-Gebirges. *Acta Biologica, Academiae Scientarum Hungaricae*, **3**, 122–124.

HOLDEN, R. 1916. A fossil wood from Burma. *Records of the Geological Survey of India*, **47**, 267–272.

HUBBARD, R. N. L. B. and BOULTER, M. C. 1983. Reconstruction of Palaeogene climate from palynological evidence. *Nature*, **301**, 147–150.

HUGHES, N. F. 1961. Fossil evidence and angiosperm ancestry. *Science Progress: A Quarterly Journal of Scientific Thought*, **69**, 84–102.

JUSSIEU, A. L. de 1789. *Genera plantarum secundum ordines naturales disposita, juxta methodum in horto regio Parisiensi exaratum, anno 1774.* Parisiis, **24**, lxxii + 498 pp.

KAISER, P. 1880. Neue fossile Laubhölzer. *Botanisches Zentralblatt*, **1**, 511–512.

KOENIGUER, J. C. 1968. Les structures ligneuses neogenes du Tchad. *Mémoires de la Section des Sciences, Comité des Travaux Historiques et Scientifiques*, **2**, 112–129.

— 1971. Sur les bois fossiles du Paléocéne de Sessao (Niger). *Review of Palaeobotany and Palynology*, **12**, 303–323.

KRAMER, K. 1974. Die Tertiaren Hölzer sudost-Asiens (unter ausschluss der Dipterocarpaceae) 1. *Palaeontographica, Abteilung B*, **144**, 45–181.

KRÄUSEL, R. 1922. Fossile Hölzer aus dem Tertiär von Süd-Sumatra. *Verhandelingen van het Geologisch-Mijnbouwkundig Genoostschap voor Nederland en Kolonien, Geologische Serie*, **5**, 231–281.

—— 1939. Ergebnisse der Foschungreisen Prof. E. Stromers in den Wüsten Ägyptens. 4. Die fossilen Floren Ägyptens. *Abhandlungen der Bayerischen Akademie der Wissenschaften, Neues Folge*, **47**, 1–140.

—— and SCHÖNFELD, G. 1924. Fossile Hölzer aus der Braunkohle von Süd-Limburg. *Abhandlungen hrsg. von den Senckenbergischen Naturforschenden Gesellschaft*, **38**, 272–282.

KRIBS, D. A. 1935. Salient lines of structural specialization in the wood rays of dicotyledons. *Botanical Gazette*, **96**, 547–557.

—— 1937. Salient lines of structural specialization in the wood parenchyma of dicotyledons. *Bulletin of the Torrey Botanical Club*, **64**, 177–186.

—— 1959. *Commercial foreign woods on the American market.* Pennsylvania, 203 pp.

KRUSE, H. O. 1954. Some Eocene dicotyledonous woods from Eden Valley, Wyoming. *Ohio Journal of Science*, **54**, 243–268.

KRYN, J. M. 1952. The anatomy of the wood of the Anacardiaceae and its bearing on the phylogeny and relationships of the family. Unpublished PhD Dissertation, University of Michigan.

KUNTH, C. S. 1824. Terebinthacearum genera denuo ad examen revocare, characteribus magis accuratis distinguere, inque septem familias distribuere conatus est. *Annales des Sciences Naturelles*, **2**, 333–366.

KURZ, S. 1874. Contributions towards a knowledge of the Burmese flora. Part 1. *Journal of the Asiatic Society of Bengal*, **43**, 39–141.

LAKHANPAL, R. N. and PRAKASH, U. 1970. Cenozoic plants from the Congo. 1. Fossil woods from the Miocene of Lake Albert. *Annales du Musée Royale de l'Afrique Centrale, Séries Sciences Géologique*, **64**, 1–20.

—— —— and AWASTHI, N. 1981. Some more dicotyledonous woods from the Tertiary of Deomali, Arunachal Pradesh, Indiaa. *Palaeobotanist*, **27**, 232–251.

LEISMAN, G. A. 1986. *Cryptocaryoxylon gippslandicum* gen. et sp. nov., from the Tertiary of eastern Victoria. *Alcheringa*, **10**, 225–234.

LIGNIER, O. 1907. Vegetaux fossiles de Normandie 4. Bois divers. *Mémoires de la Société Linnéenne de Normandie (Calvados), Premiere Série*, **22**, 237–333.

LINDLEY, J. 1830 *An introduction to the natural system of botany: or, a systematic view of the organisation, natural affinities and geographical distribution of the whole vegetable kingdom.* London, 374 pp.

—— 1836. *A natural system of botany.* London, 526 pp.

LOUVET, P. 1963. Sur un Acajou fossile du Tertiare d'Algerie: *Entandrophragmoxylon boureauii. Compte Rendu du Congrés des Sociétés Savantes de Paris et des Départments. Section de Sciences. Paris* **1963** (Moussec, 2), 493–504.

—— 1968. Sur deux Méliacées fossiles nouvelles du Tinrhert (Algerie), *Entandrophragmoxylon normandii* n. sp. et *Entandrophragmoxylon mkrattaense* n. sp. *Mémoires de la Section des Sciences, Comité des Travaux Historiques et Scientifiques*, **2**, 92–111.

—— 1970. Sur deux especes fossiles nouvelles du Lutetian superieur de Libye. *Compte Rendu du Congrés des Société Savantes de Paris et des Départments, Section des Sciences, Paris* **3**, 43–58.

—— 1971*a*. Sur la presence de *Sterculioxylon aegyptiacum* (Ung.) Kräusel et de *Leguminoxylon acaciae* Kräusel a l'ouest de l'erg oriental. *Compte Rendu du Congrés des Sociétés Savantes de Paris et des Départments. Section de, Sciences, Paris*, **94e** (Pau, 3, 1969), 115–132.

—— 1971*b*. Sur deux Léguminueses fossiles du Tinrhert (Algérie) associées à *Entandrophragmoxylon normandii* Louvet. *Compte Rendu du Congrés des Sociétés Savantes de Paris et des Départments, Section des Sciences. Paris*, **94e** (Pau, 3, 1969), 133–156.

MÄDEL, E. 1960. Mahogonihölzer der Gattung *Carapoxylon* n. g. (Meliaceae) aus dem europaischen Tertiar. *Senckenbergiana Lethaea*, **41**, 393–421.

—— 1962. Die fossilen Euphorbiaceen-Hölzer mit besonderer Berücksichtigung neuer Funde aus der Oberkreide Süd-Afrikas. *Senckenbergiana Lethaea*, **43**, 283–321.

MAHESHWARI, P. and VASIL, V. 1961. *Gnetum. Botanical Monographs* (Calcutta), **1**, vii + 142 pp.

MANCHESTER, S. R. 1977. Wood of *Tapirira* (Anacardiaceae) from the Paleogene Clarno Formation of Oregon. *Review of Palaeobotany and Palynology*, **23**, 119–127.

—— 1979. *Triplochitioxylon* (Sterculiaceae): a new genus of wood from the Eocene Clarno of Oregon and its bearing on xylem evolution in the extant genus *Triplochiton. American Journal of Botany*, **66**, 699–708.

—— 1980. *Chattawaya* (Sterculiaceae): a new genus of wood from the Eocene Clarno of Oregon and its bearing on xylem evolution in the extant genus *Pterospermum. American Journal of Botany*, **67**, 59–67.

MATHIESEN, F. J. 1932. Notes on some fossil plants from East Greenland (Cape Dalton, Turner Sound, Cape Brewster and Sabine Island). *Meddeleser om Grønland*, **85**, 5–62.

—— 1961. On two specimens of fossil wood with adhering bark from the Nugssuag Peninsula. *Meddeleser om Grønland*, **167**, 1–54.

MEIJER, J. J. F. 1997. Wood remains from the Late Cretaceous Aachen Formation. 131–137. *In* HERNGREEN, G. F. W. (ed.). *Proceedings of the 4th European Paleobotanical and Palynological Conference*. Mededelingen Nederlands Institut voor Toegepaste Geowetenschappen TNO, **58**, 302 pp.

METCALFE, C. R. 1987. *Anatomy of the dicotyledons. 3 Magnoliales, Illicales and Laurales.* Second edition. Oxford University Press, Oxford, 224 pp.

—— and CHALK, L. 1950. *Anatomy of the dicotyledons.* 1 and 2. Oxford University Press, Oxford, lxiv + 1500 pp.

MIERS, J. 1851. Observations on the affinities of the Olacaceae. *Annals and Magazine of Natural History; including Zoology, Botany and Geology, Series 2*, **8**, 161–184.

MILANEZ, F. R. 1935. Estudo de um dicotyledoneo fossil de Cretaceo. *Rodriguesia*, **1**, 83–89.

MOLL, J. W. and JANSSONIUS, H. H. 1912. The Linnean method of describing anatomical structures. Some remarks concerning the paper of Mrs. Dr. M. C. Stopes, entitled 'Petrifactions of the earliest european angiosperms'. *Verslagen van de Gewone Vergadering der Wis-en Natuurkundige Afdeeling Koninklijke Akademie van Wetenschappen te Amsterdam*, **21**, Section 15 (1), 620–629.

MUHAMMAD, A. F. and SATTLER, R. 1982. Vessel structure of *Gnetum* and the origin of angiosperms. *American Journal of Botany*, **69**, 1004–1021.

MÜLLER-STOLL, W. R. and MÄDEL, E. 1959. Betulaceen-Hölzer aus dem Tertiär des pannonischen Beckens. *Senckenbergiana Lethaea*, **40**, 159–209.

—— 1967. Die fossilen Leguminosen-Hölzer. *Palaeontographica, Abteilung B*, **119**, 95–174.

—— and MÄDEL-ANGELIEWA, E. 1984. Fossile Hölzer mit schmalen apotrachealen Parenchymbändern 3. Die Sapotaceae-Gattung *Chrysophylloxylon* gen. nov. *Feddes Repertorium*, **95**, 169–181.

MUSSA, D. 1959. Contribicao a paleoanatomia vegetal. 2. Madieras fosseis do territorio do Acre (Alto Jurua). *Boletim, Divisao de Geologia e Mineralogia*, **195**, 1–54.

NAVALE, G. K. B. 1964. *Castanoxylon* gen. nov. from Tertiary beds of the Cuddalore Series near Pondicherry, India. *Palaeobotanist*, **11**, 131–137.

NISHIDA, M., NISHIDA, H. and RANCUSI, M. 1988. Notes on petrified plants from Chile (1). *Journal of Japanese Botany*, **63**, 39–48.

PAGE, V. M. 1967. Angiosperm wood from the Upper Cretaceous of central California. 1. *American Journal of Botany*, **54**, 510–514.

—— 1968. Angiosperm wood from the Upper Cretaceous of central California. 2. *American Journal of Botany*, **55**, 168–172.

—— 1970. Angiosperm wood from the Upper Cretaceous of central California. 3. *American Journal of Botany*, **57**, 1139–1144.

—— 1979. Dicotyledonous wood from the Upper Cretaceous of central California. *Journal of the Arnold Arboretum*, **60**, 323–349.

—— 1980. Dicotyledonous wood from the Upper Cretaceous of central California. 2. *Journal of the Arnold Arboretum*, **61**, 723–748.

—— 1981. Dicotyledonous wood from the Upper Cretaceous of central California. 3. Conclusions. *Journal of the Arnold Arboretum*, **62**, 438–455.

PANT, D. D. and KIDWAI, P. F. 1971. The origin and evolution of flowering plants. *Journal of the Indian Botanical Society*, **50A**, 242–274.

PEARSON, H. L. P. 1987. Megafossil plants from Suffolk, a review of the pre-Pleistocene records. *Transactions of the Suffolk Naturalists' Society*, **23**, 56–63.

—— 1989. New records of flowering plants from Suffolk. *Transactions of the Suffolk Naturalists' Society*, **25**, 75–79.

PETRESCU, I. 1978. Studiul lemnelor fosile din Oligcenul din Nordvestul Transilvaniei. *Mémoires, Institut de Geologie et de Geophysique*, **27**, 1–84.

—— and NUTU, A. 1969. Prezenta unui lemn fosil de *Alnoxylon* in colectia muzumi din Deva. *Sargetia*, **6**, 223–229.

POITEAU, M. A. 1825. Mémoire sur les Lécythidées. *Mémoires du Muséum d'Histoire Naturelle, Paris*, **13**, 141–165.

POOLE, I. 1992. Pyritized twigs from the London Clay, Eocene, of Great Britain. *Tertiary Research*, **13**, 71–85.

—— 1993. A dipterocarpaceous twig from the Eocene London Clay Formation of southeast England. *Special Papers in Palaeontology*, **49**, 155–163.

—— 1994. "Twig" – wood anatomical characters as palaeoecological indicators. *Review of Palaeobotany and Palynology*, **81**, 33–52.

—— 1999. The presence and absence of growth ring structures in fossil 'twigs': some possible explanations. 205–219. *In* KURMANN, M. H. and HEMSLEY, A. R. (eds). *The evolution of plant architecture*. Royal Botanic Gardens, Kew, xv + 491 pp.

—— and FRANCIS, J. E. 1999. The first record of fossil atherospermataceous wood from the Upper Cretaceous of Antarctica. *Review of Palaeobotany and Palynology*, **107**, 97–107.

—— and WILKINSON, H. 1992. Two sapindaceous woods from the London Clay (Eocene) of southeast England. *Review of Palaeobotany and Palynology*, **75**, 65–75.

—— —— 1999. A celastraceous twig from the Eocene London Clay of southeast England. *Botanical Journal of the Linnean Society*, **129**, 165–176.

—— CANTRILL, D. J., HAYES, P. and FRANCIS, J. in press. The fossil record of Cunoniaceae: new evidence from Late Cretaceous fossil wood of Antarctica.

PRAKASH, U. 1968. Miocene fossil wood from the Columbia basalts of central Washington. 3. *Palaeontographica, Abteilung B*, **96**, 66–97.

—— 1976. Fossil woods resembling *Dichrostachys* and *Entandrophragma* from the Tertiary of the Middle East. *Abhandlungen des Zentralen Geologischen Instituts*, **26**, 499–507.

—— 1978. Fossil woods from the Lower Siwalik Beds of Uttar Pradesh India. *Palaeobotanist*, **25**, 378–392.

—— and BANDE, M. B. 1980. Some more fossil woods from the Tertiary of Burma. *Palaeobotanist*, **26**, 261–278.

—— —— and LALITHA, V. 1986. The genus *Phyllanthus* from the Tertiary of India with critical remarks on the nomenclature of fossil Euphorbiaceae. *Palaeobotanist*, **35**, 106–114.

—— BREZINOVA, D. and AWASTHI, N. 1974. Fossil woods from the Tertiary of South Bohemia. *Palaeontographica, Abteilung B*, **147**, 107–123.

—— —— and BUŽEK, C. 1971. Fossil woods from the Doupovske Horý and Ceské Stredohori mountains in northern Bohemia. *Palaeontographica, Abteilung B*, **133**, 103–128.

—— and DAYAL, R. 1965. *Barringtonioxylon eopterocarpum* sp. nov., a fossil wood of Lecythidaceae from the Deccan Intertrappean beds of Mahurzari. *Palaeobotanist*, **13**, 25–29.

—— and TRIPATHI, P. P. 1969. Fossil woods of Leguminosae and Anacardiaceae from the Tertiary of Assam. *Palaeobotanist*, **17**, 22–32.

—— —— 1972. Fossil woods of *Careya* and *Barringtonia* from the Tertiary of Assam. *Palaeobotanist*, **19**, 55–160.

—— —— 1974. Fossil woods from the Tertiary of Assam. *Palaeobotanist*, **21**, 305–316.

—— —— 1975. Fossil dicotyledonous woods from the Tertiary of eastern India. *Palaeobotanist*, **22**, 51–62.

PRESTWICH, J. 1852. On the structure of the strata between the London Clay and the Chalk in the London and Hampshire Tertiary systems. Part 3. The Thanet Sand Formation. *Quarterly Journal of the Geological Society of London*, **8**, 235–268.

PRIVÉ, C. 1973. *Tetrapleuroxylon limagnense* n. sp. bois fossile de L'Allier (France). *Review of Palaeobotany and Palynology*, **15**, 301–313.

RECORD, S. J. and HESS, R. W. 1943. *Timbers of the New World.* New Haven and London, xi + 640 pp.

REID, E. M. and CHANDLER, M. E. J. 1933. *The flora of the London Clay.* The Natural History Museum, London, viii + 561 pp.

RIBBINS, M. M. and COLLINSON, M. E. C. 1978. Further notes on pyritised fern rachides from the London Clay. *Tertiary Research*, **2**, 47–50.

RICHTER, H. G. 1987. Lauraceae, mature secondary xylem. 162–169. *In* METCALFE, C. R. *Anatomy of the Dicotyledons. 3. Magnoliales, Illicales and Laurales.* Second edition. Oxford University Press, Oxford, 224 pp.

SALARD, M. 1963. Sur un bois tertiare du Pérou. *Compte Rendu du Congrés des Sociétés Savantes de Paris et des Départments, Section des Sciences. Paris*, **88e** (Moussec), 483–492.

SCHENK, A. 1883. Fossile Hölzer. *Palaeontographica, Abteilung B*, **30**, 1–17.

SCHILKINA, I. A. 1956. *In* KIPARIAOVA, L. S., MARKOVSKI, B. P. and RADCHENKO, G. P. (eds). Novye semeistva i rody, materialy po paleontologii. *Trudy Vsesoyuznogo Naucho-issledovatel'skogo Geologicheskogo Instituta (VSEGEI), Novaya Seriya*, **12**, 260 pp.

—— 1973. A new species of *Anacardioxylon* from the Tertiary deposits of eastern Transcaucasia. *Botanicheskiy Zhurnal, SSSR*, **56**, 827–829.

SCHLEIDEN, M. J. 1853. Il Ueber die fossilen Pflanzenreste des Jenaischen Muschelkalks, 9–30. *In* SCHMIDT, E. Die organischen Reste des Muschelkalkes im Saalthales bei Jena. *Neues Jahrbuch für Mineralogie, Geognosie, Geologie und Petrefactekunde*, **1853**, 1–72.

SCHÖNFELD, E. 1956. Fossile Hölzer von Island. *Neues Jahrbuch für Geologie und Paläontologie, Abhandlungen*, **104**, 191–225.

SCHÖNFELD, G. 1947. Hölzer aus dem Tertiar von Kolumbien. *Abhandlungen Herausgegeben von der Senckenbergischen Naturforschenden Gesellschaft*, **475**, 1–53.

SCHUSTER, J. 1911. Monographie der fossilien Flora der *Pithecanthropus*-Schichten. *Abhandlungen der Bayerischen Akademie der Wissenschaften*, **25**, 1–70.

SCHWEITZER, H.-J. 1958. Die fossilen dipterocarpaceen-Hölzer. *Palaeontographica, Abteilung B*, **105**, 1–66.

SCOTT, A. C. and DE KLERK, R. 1974. A preliminary study of London Clay pyritized 'twigs' from the Isle of Sheppey. *Tertiary Research*, **2**, 73–82.

SCOTT, D. H. 1924. *Extinct plants and problems of evolution.* Macmillan, London, xiv + 240 pp.

SCOTT, R. A. and WHEELER, E. A. 1982. Fossil woods from the Eocene Clarno Formation of Oregon. *Bulletin of the International Association of Wood Anatomists, New Series*, **3**, 135–154.

SELMEIER, A. 1970. Ein verkieseltes Ilex-Holz, *Ilicoxylon austriacum* n. sp., aus den Atzbacher Sanden (Ottnangien) von Gallspach. *Neues Jahrbuch für Geologie und Paläontologie, Monatshefte*, **1970**, 683–700.

—— 1975. *Fagaroxylon bavaricum* n. sp., ein verkieseltes Rutaceen-Holz aus Obermiozänen Schicten Niederbayerns. *Mitteilungen der Bayerischen Staatssammlung für Paläontologie und Historische Geologie*, **15**, 157–167.

—— 1990. Ein verkieseltes Juglandaceen-Holz aus den Elbe-Geröllen von Hoyerswerda (Brandenburg). *Mitteilungen der Bayerischen Staatssammlung für Paläontologie und Historische Geologie*, **30**, 107–119.

—— 1991. Ein verkieseltes Sapotaceae-Holz, *Bumelioxylon holleisii* n. gen., n. sp., aus jungtertiären Schichten der Südlichen Frankenalb (Bayern). *Archaeopteryx*, **9**, 55–72.

SERLIN, B. S. 1982. An Early Cretaceous fossil flora from northwest Texas, its composition and implications. *Palaeontographica, Abteilung B*, **182**, 52–86.

SEWARD, A. C. 1919. *Fossil plants. 4.* Cambridge University Press, Cambridge, xvi + 543 pp.

—— and HOLLTUM, R. E. 1924. Tertiary plants from Mull. 67–70. *In* BAILEY, E. B., CLOUGH, C. T., WRIGHT, W. B., RICHEY, J. E. and WILSON, G. V. (eds). *Tertiary and post-Tertiary geology of Mull, Loch Aline and Oban.* Memoirs of the Geological Survey of Great Britain.

SHACKLETON, N. J. 1986. Palaeogene stable isotope events. *Palaeogeography, Palaeoclimatology, Palaeoecology*, **57**, 91–102.

SHALLOM, J. 1960. Fossil dicotyledonous wood of Lecythidaceae from the Deccan Intertrappean beds of Mahurzari. *Journal of the Indian Botanical Society*, **39**, 198–203.

SHAVROV, L. A. 1956. The wood anatomy of the *Ephedra* species of the U.S.S.R. *Botanicheskiy Zhurnal, SSSR*, **41**, 1324–1331.

SPACKMAN, W. 1948. A dicotyledonous wood found associated with the Idaho tempskyas. *Annals of the Missouri Botanical Garden*, **35**, 107–115.

STOPES, M. C. 1912. Petrifactions of the earliest European angiosperms. *Philosophical Transactions of the Royal Society, Series B*, **203**, 75–100.

—— 1915. *The Cretaceous flora, Pt. 2. Lower Greensand (Aptian) plants of Britain.* Catalogue of Mesozoic Plants in The Natural History Museum, British Museum (Natural History), London, 360 pp.

—— and FUJII, K. 1910. Studies on the structure and affinities of Cretaceous plants. *Philosophical Transactions of the Royal Society, Series B*, **201**, 1–90.

SÜSS, H. 1960. Ein Monimiaceen-Holz aus der oberen Kreide Deutschlands, *Hedycaryoxylon subaffine* (Vater) nov. comb. *Senckenbergiana Lethaea*, **41**, 317–330.

—— 1986. Untersuchungen über fossile Buchenhölzer Beiträge zu einer Monographie der Gattung *Fagoxylon* Stopes and Fujii. *Feddes Repertorium*, **97**, 161–183.

—— and MÜLLER-STOLL, W. R. 1977. Untersuchungen gen über fossile Platenhölzer. Beitrage zu einer Monographie der Gatttung *Platanoxylon* Andreanszky. *Feddes Repertorium* **88**, 1–62.

THAYN, G. F., TIDWELL, W. D. and STOKES, W. L. 1983. Flora of the Lower Cretaceous Cedar Mountain Formation of Utah and Colorado. Part 1. *Paraphyllanthoxylon utahense. Great Basin Naturalist*, **43**, 394–402.

—— 1985. Flora of the Lower Cretaceous Cedar Mountain Formation of Utah and Colorado. Part III: *Icacinoxylon pittiense* n. sp. *American Journal of Botany*, **72**, 175–180.

TOMKIEFF, S. J. and BLACKBURN, K. B. 1942. On the remains of fossil wood enclosed in a Tertiary lava on the Isle of Rhum, Inner Hebrides. *Geological Magazine*, **79**, 14–17.

TORRES, T. and LEMOIGNE, Y. 1989. Hallazagos de maderas fósiles de Angiospermas y Gimnospermas del Cretácico Superior en punta Williams, isla Livingston, islas Shetland del Sur, Antártica. *Serie Cientas INACH*, **39**, 9–29.

TUCKEY, J. H. 1818. *Narrative of an expedition to explore the river Zaire, usually called the Congo, in South Africa, in 1816: to which is added the journal of Prof. [C.] Smith, some general observations on the country and its inhabitants: and an appendix containing the natural history of that part of the Kingdom of Congo through which the Zaire flows.* John Murray, London, lxxxii + 498 pp., 13 pls, 1 map.

UNGER, F. 1842. *Synopsis lignorum fossilium plantarum acramphibryarum. In* ENDLICHER, S. (ed.). *Genera plantarum, Supplement* **2**, 100–102. Mantissa.

—— 1845. *Synopsis plantarum fossilium.* Leipzig, 330 pp.

VAN VLIET, G. J. C. M. 1979. Wood anatomy of the Combretaceae. *Blumea*, **25**, 141–223.

WARD, D. W. 1978. The lower London Tertiary (Palaeocene) succession of Herne Bay, Kent. *Report of the Institute of Geological Sciences*, **78**, 10 pp.

WHEELER, E. A. 1986. Vessels per square millimetre or vessel groups per square millimetre? *Bulletin of the International Association of Wood Anatomists, New Series*, **7**, 73–74.

—— 1991. Paleocene dicotyledonous trees from Big Bend National Park, Texas; variability in wood types common in the Late Cretaceous and early Tertiary and ecological inferences. *American Journal of Botany*, **78**, 658–761.

—— and BAAS, P. 1991. A survey of the fossil record for dicotyledonous wood and its significance for evolutionary and ecological wood anatomy. *Bulletin of the International Association of Wood Anatomists, New Series*, **12**, 275–332.

—— —— 1992. Fossil wood of the Leguminosae: a case study in xylem evolution and ecological anatomy. 281–301. *In* HERENDEEN, P. S. and DILCHER, D. L. (eds). *Advances in legume systematics: 4. The fossil record.* Royal Botanic Gardens, Kew, 326 pp.

—— —— 1993. The potential and limitations of dicotyledonous wood anatomy for climatic reconstructions. *Paleobiology*, **19**, 487–498.

—— —— and GASSON, P. E. (eds) 1989. IAWA list of microscopic features for hardwood identification. With an appendix on non-anatomical information. *Bulletin of the International Association of Wood Anatomists, New Series*, **10**, 219–332.

—— LEHMAN, T. M. and GASSON, P. E. 1994. *Javelinioxylon*, an Upper Cretaceous dicotyledonous tree from Big Bend National Park, Texas, with presumed malvalean affinities. *American Journal of Botany*, **81**, 703–710.

—— McCLAMMER, J. and LAPASHA, C. A. 1995. Similarities and differences in dicotyledonous woods of the Cretaceous and Palaeocene, San Juan Basin, New Mexico, USA. *Bulletin of the International Association of Wood Anatomists, New Series*, **16**, 223–254.

—— LEE, M. and MATTEN, L. C. 1987. Dicotyledonous woods from the Upper Cretaceous of southern Illinois. *Botanical Journal of the Linnaean Society*, **95**, 77–100.

—— PEARSON, R. G., LAPASHA, C. A., ZACK, T. and HATLEY, W. 1986. Computer-aided wood identification. *Bulletin of the North Carolina Agricultural Research Station*, **474**, 1–60.

—— SCOTT, R. A. and BARGHOORN, E. S. 1977. Fossil dicotyledonous woods from Yellowstone National Park. *Journal of the Arnold Arboretum*, **58**, 280–302.

—— —— —— 1978. Fossil dicotyledonous woods from Yellowstone National Park, II. *Journal of the Arnold Arboretum*, **59**, 1–26.

WIEMANN, M. C., MANCHESTER, S. R. and WHEELER, E. A. 1999. Paleotemperature estimation from dicotyledonous wood anatomical characters. *Palaios*, **14**, 459–474.

—— WHEELER, E. A., MANCHESTER, S. R. and PORTIER, K. M. 1998. Dicotyledonous wood anatomical characters as predictors of climate. *Palaeogeography, Palaeoclimatology, Palaeoecology*, **139**, 83–100.

WILKINSON, H. P. 1981. The anatomy of the hypocotyls of *Ceriops* Arnott (Rhizophoraceae) Recent and fossil. *Botanical Journal of the Linnean Society*, **82**, 139–164.

—— 1983. Starch grain casts and moulds in Eocene (Tertiary) fossil mangrove hypocotyls. *Annals of Botany*, **51**, 39–45.

—— 1984. Pyritized twigs from Sheppey. *Tertiary Research*, **5**, 189–198.

—— 1988. Sapindaceous pyritized twigs from the Eocene of Sheppey, England. *Tertiary Research*, **9**, 81–86.

WITT, H. C. D. de 1966. *Plants of the world. The higher plants. 1.* Thames and Hudson, London, 335 pp.

WOLFE, J. A. 1979. Temperature parameters of humid to mesic forests of eastern Asia and relation to forests of other regions of the Northern Hemisphere and Australia. *United States Geological Survey, Professional Paper*, **1106**, 1–37.

—— 1985. Distribution of major vegetational types during the Tertiary. 357–375. *In* SUNDQUIST, E. T. and BROECKER, W. S. (eds). *The carbon cycle and atmospheric CO2: natural variation Archean to Present.* Geophysical Monographs, **32**, 375 pp.

—— and UPCHURCH, G. R. 1987. North American nonmarine climates and vegetation during the Late Cretaceous. *Palaeogeography, Palaeoclimatology, Palaeoecology*, **61**, 33–77.

WORBES, M. 1989. Growth rings, increment and age of trees in inundation forests, savannas and a mountain forest in the neotropics. *Bulletin of the International Association of Wood Anatomists, New Series*, **10**, 109–122.

WORSSAM, B. C. 1963. *Geology of the country around Maidstone.* HMSO, London, viii + 152 pp.

—— 1973. A new look at river capture and at the denudation history of the Weald. *Report of the Institute of Geological Sciences*, **73** (17), 21 pp.

MARK CRAWLEY

Department of Palaeontology
The Natural History Museum
London SW7 5BD, UK

Current address
201 Charterhouse Road
Orpington
Kent BR6 9ET, UK